Jeff Evans

NEW PERSPECTIVES ON NITROGEN CYCLING IN THE TEMPERATE AND TROPICAL AMERICAS

NEW PERSPECTIVES ON NITROGEN CYCLING IN THE TEMPERATE AND TROPICAL AMERICAS

Report of the International SCOPE Nitrogen Project

Edited by

ALAN R. TOWNSEND

University of Colorado, Boulder, Colorado, U.S.A.

Reprinted from *Biogeochemistry*

Volume 46, Nos. 1–3 (1999)

KLUWER ACADEMIC PUBLISHERS

DORDRECHT/BOSTON/LONDON

A C.I.P. catalogue record for this book is available from the Library of Congress.

ISBN 0-7923-5704-3

Published by Kluwer Academic Publishers,
P.O. Box 17, 3300 AA Dordrecht, The Netherlands.

Sold and distributed in the U.S.A. and Canada
by Kluwer Academic Publishers,
101 Philip Drive, Norwell, MA 02061, U.S.A.

In all other countries, sold and distributed
by Kluwer Academic Publishers,
P.O. Box 322, 3300 AH Dordrecht, The Netherlands.

Printed on acid-free paper.

Printed in the Netherlands.

Table of Contents

Biogeochemistry **46:** 1–3, 1999.

Foreword

The papers in this special issue grew out of a workshop held in Termás de Chillán, Chile, in December of 1996. This was the fourth of a series of workshops that comprise the core activities of the International SCOPE Project on Nitrogen Transport and Transformations: A Regional and Global Analysis; the first of these workshops also resulted in a special issue of this journal (*Biogeochemistry* 35(1): 1996). The Chilean workshop's central purposes were to compare nitrogen cycling in the relatively pristine temperate South Americas with the generally more polluted temperate North Americas, and to compare both with nitrogen cycling in the tropical Americas. More than 40 scientists from 12 different countries helped make this meeting a success, and their enthusiasm for the ideas generated during the meeting carried over into the production of the twelve manuscripts included in this volume. It is my belief that these manuscripts are rich in exciting new ideas and theory; I hope the readers of *Biogeochemistry* share this opinion.

The first contribution in this issue is by Jorge Corredor; he provides a brief but enlightening and appropriate history of the word nitrogen. As he points out, many of us use nitrogen and related words on a near-daily basis, yet few of us know their etymology. The next paper is by Elisabeth Holland and colleagues; they synthesize data on N deposition and N emissions as well as use a global model of tropospheric chemistry and transport (MOGUNTIA) to provide an updated view of reactive N exchanges between the terrestrial biosphere and the atmosphere.

The next nine papers can be thought of as three somewhat distinct sections of this volume. The first three of these nine all focus on N cycling in terrestrial ecosystems, with an emphasis on the moist tropics. Luiz Martinelli et al. present data on ^{15}N values of both foliage and soils from a variety of tropical and temperate forests, including forests along a soil age gradient in the Hawaiian Islands. They show substantial differences in ^{15}N patterns between temperate and tropical systems, and argue that these isotopic data provide time-integrated evidence that N cycles in relative excess in many tropical forests. Pamela Matson and colleagues then present a theoretical analysis of how moist tropical forests may respond to increasing N deposition in the next few decades; they contend that the naturally N-rich state of these systems will lead to substantially different responses to increasing N additions than has been seen in the temperate zone. Finally, Tamara Chestnut and co-authors present a nitrogen budget for hillslope tabonuco forest stands in Puerto Rico; such budgets for tropical systems are both rare and sorely needed.

The next three papers address N cycling in aquatic ecosystems, again with a focus on the tropics. John Downing and the other members of the aquatic working group from Chillán examine how land use and land cover changes in the tropics may affect N cycling in both freshwater and coastal marine ecosystems. They conclude their analyses with the hypothesis that alterations to the nitrogen cycle will affect tropical aquatic systems to an even greater extent than what has already been seen in the temperate zone. Bill Lewis and co-authors then summarize N export from thirty-one relatively pristine watersheds, thereby providing valuable information on the baseline conditions against which we can assess anthropogenic effects on the N cycle. They focus on the tropics, but present data from a few temperate watersheds as well, and show that N yields and concentrations in the tropical systems are much greater than those seen in the temperate watersheds. Jorge Corredor et al. conclude this group of papers with a look at nutrient limitation and its response to human activities in tropical marine systems, with a focus on the Caribbean.

The next three papers all tackle an enduring paradox in ecosystem ecology: the question of why N limitation remains so widespread despite the occurrence of organisms that can fix atmospheric nitrogen. Eight years ago in this journal, Peter Vitousek and Bob Howarth published a widely read and influential analysis of this issue. Both scientists (along with co-authors) revisit the question here. Vitousek and Chris Field present a new simulation model of N fixation in terrestrial systems, one in which costs of fixation, N losses, light availability, P limitation and grazing are all addressed. They use this model to explore the potential constraints on fixation in the terrestrial biosphere, as well as how both N fixation and limitation may respond to elevated CO_2. Howarth et al. also describe a new model of N fixation, this time one designed for cyanobacteria in estuaries. Their conclusions, namely that a combination of biotic (especially grazing) and biogeochemical (notably Mo availability) constraints can interact to keep rates of N fixation low to absent in estuarine sytems, have striking parallels to the Vitousek and Field paper, and suggest that more general, unifying models of N fixation may be possible. Finally, Tim Crews concludes this section with an alternative view on the patterns in symbiotic N fixation that occur in the terrestrial biosphere. Crews argues that the presence or absence of N-fixing plant symbioses should first be viewed in an evolutionary context, with a focus on the radiation of the family *Fabaceae*.

David Hooper and Loretta Johnson conclude the issue with an investigation of the interactions between nitrogen limitation and water availability in dryland ecosystems. They survey published data on responses to fertilizer across a broad-scale gradient in annual precipitation, and conclude that there is not a shift in primary limitation from water to N across this gradient, but that co-limitation better explains the data.

The Termás de Chillán workshop was sponsored by the Mellon Foundation and the Inter-American Institute for Global Change (IAI). The Mellon Foundation also supports the SCOPE Nitrogen Project. Bob Howarth and I served as co-chairs for the meeting; other members of the steering committee included Peter Vitousek, Doris Soto, Juan Armesto and Osvaldo Sala. Juan Armesto was the local host and suggested the wonderful location in which the workshop was held. The boundless energy and enthusiasm shown by participants both during and after the meeting were instrumental to the success of this volume; my sincere thanks to all of you.

Alan R. Townsend, Editor
University of Colorado, U.S.A.

International Scope Project on Nitrogen Transport and Transformations: A Regional and Global Analysis

Objectives and Activities

- to foster the necessary synergism between scientist of many disciplines (marine ecologists, forest ecologists, agricultural scientists, microbiologists, atmospheric chemists, oceanographers, hydrologists) in order to help develop new approaches for the study of nitrogen cycling;
- to refine the global nitrogen budget and develop regional budgets for selected key and contrasting regions of the world;
- to more fully understand the problems stemming from accelerated nitrogen cycling, and the inter-relationships among these problems.

Scientific Advisory Committee

Co-chairs:
Robert Howarth (USA)
John Freney (Australia)

Members:
Frank Berendse (The Netherlands)
Pornpimol Chaiwanakupt (Thailand)
Valery Kudeyarov (Russia)
Scott Nixon (USA)
Peter Vitousek (USA)
Zhu Zhao-liang (People's Republic of China)

Consultants:
Ragner Elmgren (Sweden)
James Galloway (USA)
Alan Townsend (USA)

Biogeochemistry **46:** 5–6, 1999.

Νιτρον – An etymology of nitrogen and other related words

JORGE E. CORREDOR
University of Puerto Rico, Department of Marine Sciences, Mayaguez, PR 00680, Puerto Rico

Received 10 December 1998

Nitrogen, ammonium, diazo salts and *diazotroph* are words in common use by scientists, yet few of those concerned with nitrogen research are aware of their etymology. This note is intended to address the issue.

Perhaps the most ancient word related to nitrogen is "nitre," the name for its sodium salt known in modern times as sodium nitrate and in earlier days as salt peter. Most dictionaries trace this word from the Greek "*νιτρον*" through the latinized form "*nitrum*". Some, however, would ascribe a more ancient origin to the root arising from the Egyptian and rendered as "*ntry*".

The word "*ammonium*" has a similarly remote etymological origin. During the final centuries of the Roman empire, Greco-Roman influence had extended throughout the Mediterranean region, reaching as far as Egypt and Libya. This influence was so pervasive that the major Greco-Roman and Egyptian gods were, in some cases, fused into one deity, as was the case for Jupiter-Ammon, the king of Greco-Roman gods; and, as Amen Ra, king of gods and god of the sun for the Egyptians. At this time, there existed a temple dedicated to the god in Lybia where "ammoniacal" salts were first extracted. Although there are different versions as to the precise extraction process, both have to do with camels. Mendeleeff (1905) states that the salts were extracted from the soot accumulated from the burning of camel dung, while Webster's Dictionary of Word Origins (1991) holds that camel urine from a cesspool near the temple was heated with soot and seasalt to form sal ammoniac; the "salt of Ammon". Today, we use the root not only to designate the salts, but also in words which relate to compounds containing the parent gas *ammonia* and its organic derivatives, the *amino* compounds.

Until the eighteenth century, air was considered to be a single gas. At this time, the experiments of Priestly and, more forcibly, those of Lavoisier, demonstrated that, upon combustion, a fraction of the air was consumed, but a much larger fraction remained unaffected by combustion (the original discovery of the gas is, however, attributed to Scheele in 1772, who distinguished between "foul air" and "fire air," and to Rutherford concurrently or soon thereafter). The gas remaining after combustion was, in any case, shown by Lavoisier to be incapable of supporting life, hence being termed the "*a-zooic*" gas. Lavoisier's original terminology persists in the French name for the gas, "*azote*," and in English and other languages in words such as "*diazotroph*", an organism capable of consuming molecular nitrogen (a compound word using the French diazo root combined with the Greek to signify "eater of nitrogen"), and "*diazo*" compounds, those incorporating two double-bonded nitrogen molecules.

The word "*nitrogen*" itself was first coined in 1790 by Chaptal and remains the word of preference for naming the gas in many of the romance languages. The word arises from the Greek roots for "*generator of nitre*".

References

Asimov I (1965) A Short History of Chemistry. Doubleday-Anchor, N.Y.

Mendeleeff D (1902) The Principles of Chemistry, Vol I (p 247). American Home Library Company

Webster's Dictionary of Word Origins (1991) Smithmark, N.Y.

Biogeochemistry **46:** 7–43, 1999.

Contemporary and pre-industrial global reactive nitrogen budgets

ELISABETH A. HOLLAND[1,2], FRANK J. DENTENER[3], BOBBY H. BRASWELL[4] & JAMES M. SULZMAN[1]
[1]*Atmospheric Chemistry Division, National Center for Atmospheric Research, Boulder, CO 80307 U.S.A.;* [2]*Max-Planck-Institut für Biogeochemie, D-00745 Jena, Federal Republic of Germany;* [3]*Utrecht University, Institute for Marine and Atmospheric Research (IMAU), Princetonplein 5, NL-3584 CC Utrecht, The Netherlands;* [4]*Institute for the Study of Earth, Oceans and Space, University of New Hampshire, Durham, NH 03824 U.S.A.*

Received 10 December 1998

Key words: ammonia emissions, global nitrogen cycle, nitric oxide, nitrogen deposition, nitrogen pollution

Abstract. Increases and expansion of anthropogenic emissions of both oxidized nitrogen compounds, NO_x, and a reduced nitrogen compound, NH_3, have driven an increase in nitrogen deposition. We estimate global NO_x and NH_3 emissions and use a model of the global troposphere, MOGUNTIA, to examine the pre-industrial and contemporary quantities and spatial patterns of wet and dry NO_y and NH_x deposition. Pre-industrial wet plus dry NO_x and NH_x deposition was greatest for tropical ecosystems, related to soil emissions, biomass burning and lightning emissions. Contemporary NO_y + NH_x wet and dry deposition onto Northern Hemisphere (NH) temperate ecosystems averages more than four times that of pre-industrial N deposition and far exceeds contemporary tropical N deposition. All temperate and tropical biomes receive more N via deposition today than pre-industrially. Comparison of contemporary wet deposition model estimates to measurements of wet deposition reveal that modeled and measured wet deposition for both NO_3^- and NH_4^+ were quite similar over the U.S. Over Western Europe, the model tended to underestimate wet deposition of NO_3^- and NH_4^+ but bulk deposition measurements were comparable to modeled total deposition. For the U.S. and Western Europe, we also estimated N emission and deposition budgets. In the U.S., estimated emissions exceed interpolated total deposition by 3–6 Tg N, suggesting that substantial N is transported offshore and/or the remote and rural location of the sites may fail to capture the deposition of urban emissions. In Europe, by contrast, interpolated total N deposition balances estimated emissions within the uncertainty of each.

Abbreviations: EMEP – European Monitoring and Evaluation Program; GEIA – Global Emissions Inventory Activity; NADP/NTN – National Atmospheric Deposition Program/National Trends Network in the U.S.; NH – Northern Hemisphere; $NH_x = NH_3 + NH_4^+$; $NO_x = NO + NO_2$; NO_y – total odd nitrogen = $NO_x + HNO_3 + HONO + HO_2NO_2 + NO_3$ + radical $(NO_3^{\cdot})$+; Peroxyacetyl nitrates + N_2O_5 + organic nitrates; SH – Southern Hemisphere; Gg – 10^9 g; Tg – 10^{12} g

Introduction

Inputs of nitrogen to terrestrial and aquatic ecosystems have increased several-fold over the last one hundred and fifty years, with the steepest increases during the last four decades. The expansion of fertilizer manufacture and use, the increase in fossil fuel combustion, the intensification of animal husbandry, and widespread cultivation of N_2 fixing crops have all contributed to the dramatic increase in N inputs. The increase has been most rapid in Northern Hemisphere (NH) temperate ecosystems, but presently subtropical and tropical regions of Asia are also experiencing an explosive increase in N inputs to terrestrial ecosystems (W. Chameides, pers. comm.; Galloway et al. 1996). Projected increases in N deposition for these tropical and subtropical regions, with a high natural background of N inputs, exceed increases projected for temperate and arctic regions (Cleveland et al. submitted; Galloway et al. 1994; Holland & Lamarque 1997a). Compared to biological N fixation, N deposition is becoming a proportionately greater source of N to terrestrial and aquatic ecosystems worldwide (Vitousek et al. 1997).

The nitrogen contained in the atmosphere as N_2, $3.9 * 10^6$ Tg (Tg = 10^{12} g), is the largest reservoir of N in the Earth system (Warneck 1988). However, this paper focuses on the nitrogen emissions and deposition that have been transformed from N_2 into reactive forms that are biologically available (e.g. Vitousek et al. 1997). We consider the emissions of NO_x (NO + NO_2) and ammonia (NH_3) as well as their respective deposition products: NH_x (NH_3 + NH_4^+) and NO_y (total odd nitrogen, NO_x + HNO_3 + HONO + HO_2NO_2 + $RC(O)O_2NO_2$ + $NO_3^{\cdot}$ + N_2O_5 + $RONO_2$, where R is a higher alkyl group) (Ridley et al. 1996). Although deposition of all of these chemical species can be modeled, the deposition products most frequently measured are deposition of nitrate, NO_3^-, and ammonium, NH_4^+, in precipitation. Nitrate in rainwater is most commonly the dissolution product of nitric acid (HNO_3), but may also be the product of the dissolution of salts (e.g. $Ca(NO_3)_2$, $NaNO_3$, and NH_4NO_3), as well as organic nitrates. Ammonium is the dissolution product of compounds such as $(NH_4)_2SO_4$ and NH_4NO_3. There is increasing evidence that deposition of organic N may be important but there are few measurements (Cornell et al. 1995; Elkund et al. 1997; Rendell et al. 1993). Since there is little chemical interconversion between NO_x and NH_x in the atmosphere, on a global basis deposition of oxidized nitrogen (NO_y) and ammonia containing compounds (NH_x) will balance their respective emissions according to the law of mass conservation. The nitrogen-containing compound emitted (e.g. NO_x & NH_3) is not always the same as the compound deposited (e.g. HNO_3 and NH_x), but the nitrogen itself is conserved. The aim of this paper is to examine the patterns, sources and magnitudes of pre-industrial and contemporary N deposition onto different ecosystem types.

Methods

To begin to understand how humans have changed N inputs to terrestrial ecosystems, we first used extant data to provide an up-to-date estimate of the global inventory of NO_x and NH_3 emissions. We then used a model of the troposphere, MOGUNTIA, to estimate N deposition inputs for both pre-industrial and modern scenarios (Dentener & Crutzen 1993; Dentener & Crutzen 1994). MOGUNTIA is currently the only global 3-D chemical transport model that simulates both NO_x emissions and its deposition (wet & dry) as NO_y, as well as NH_3 emissions and its deposition (wet & dry) as NH_x (Dentener & Crutzen 1994). In a previous study, MOGUNTIA-simulated NO_y deposition was similar to other models ranging in resolution from 2.4° by 2.4° up to the coarse resolution of 10° by 10° (Holland et al. 1997b). We then examined which vegetation types receive the wet and dry deposited N based on pre-industrial and contemporary vegetation distributions. Relatively clean Southern Hemisphere (SH) temperate ecosystems contrast strongly with NH temperate ecosystems, where human influence on N emissions and deposition is the greatest, and with tropical ecosystems where biological inputs and natural emissions of N containing trace gases are greatest.

Emission estimates

We reviewed the literature to construct global estimates of pre-industrial and modern emissions of the two most important classes of N emissions: NO_x and NH_3. The purpose of this exercise was: (1) provide a background against which to interpret the emissions used in the MOGUNTIA simulations, and (2) incorporate these newest estimates into the budget. A large number of global compilations of natural and anthropogenic NO_x-N emissions have been published recently (Benkovitz et al. 1996; Davidson & Kingerlee 1997; Delmas et al. 1997; Galloway et al. 1995; Holland et al. 1997b; Lee et al. 1997; Logan 1983; Prather et al. 1995; Price et al. 1997a; Price et al. 1997b; Sanhueza et al. 1995; Whelpdale et al. 1997). Two of the compilations, one of soil NO_x emissions and the other of global NO_x emissions, were the direct result of the SCOPE activity convened in Tsukuba, Japan in the spring of 1996 (Davidson & Kingerlee 1997; Delmas et al. 1997). A somewhat smaller number of contemporary global compilations of NH_3-N emissions have been published (Bottger et al. 1978; Bouwman et al. 1997; Dentener & Crutzen 1994; Galloway et al. 1995; Schlesinger & Hartley 1992; Soderlund

& Svensson 1976; Stedman & Shetter 1983; Warneck 1988). The most recent contemporary NH_3 emission inventory (Bouwman et al. 1997) is the most comprehensive.

Deposition estimates

Global estimates of total N deposition onto land were made using a model of the troposphere, MOGUNTIA, described by Zimmerman (1987) and modified by Dentener and Crutzen (1993, 1994). The modifications enable large scale modeling of atmospheric cycles of oxidized N: NO_x and NO_y, and reduced N: NH_3 and ammonium sulfate, $(NH_4)_2SO_4$. Like other 3-D chemical transport models, MOGUNTIA incorporates emissions of NO_x-N and NH_x-N (the latter is unique to MOGUNTIA) which are released on a latitude by longitude grid and transported. The sources of NH_3 and NO_x used in both the pre-industrial and contemporary runs are described in Tables 1 and 2. The emitted compounds undergo chemical transformation, and are then either deposited back to the surface as wet or dry deposition. We do not include surface emissions of nitrous oxide (N_2O), the largest source of surface-emitted N to the stratosphere, because it is unreactive in the troposphere. The primary mechanisms for NO_x-N, NO_y-N, and NH_x-N removal from the atmosphere are through wet deposition (via precipitation) and dry deposition of gases and particulates.

MOGUNTIA has a relatively coarse resolution of 10° by 10° grids; the modeled atmosphere extends from the surface to approximately 16 km and is divided into 10 layers of 100 hPa thickness (Zimmermann 1987). The model is driven using monthly average winds and temperatures from Oort (1983) and precipitation climatologies from Jaeger (1976). The time step of the photochemical portion of the model is two hours. Important features of the transport scheme include deep cumulus convection (Feichter & Crutzen 1990) and eddy diffusion based on the standard deviation of the monthly mean winds. The chemical species transported include NO_x, HNO_3, O_3, CO, CH_4, H_2O_2, H^+, SO_4^{2-}, NH_3, NH_4^+, CH_2O, DMS, SO_2, C_2-C_3^-, and peroxy-acetyl nitrate (PAN). The chemical scheme includes CH_4-CO-NO_x and HO_x with the associated photochemistry, heterogeneous destruction of N_2O_5, and the integration of sulfur and NH_x chemistry. Dry deposition is calculated according to

$$F = V_d n Y, \qquad (1)$$

where V_d is the deposition velocity, n is molecules cm^{-3}, and Y is the mixing ratio of the gas. The deposition velocities are specific to the chemical species considered and vary depending on surface cover. The deposition velocities

Table 1. Global emissions of NO_x in Tg N y^{-1} to the troposphere for the mid 1800s and the 1980–90s compiled from a literature review (Benkovitz et al. 1996; Davidson & Kingerlee 1997; Delmas et al. 1997; Galloway et al. 1994; Holland et al. 1997b; Lee et al. 1997; Logan 1983; Prather et al. 1995; Price et al. 1997a; Sanhueza et al. 1995; Whelpdale et al. 1997). Values in parentheses are those used in the MOGUNTIA simulations described below (Dentener & Crutzen 1993; Dentener & Crutzen 1994).

Source	Pre-industrial	Contemporary
Fossil fuel combustion	0	20–24 (20)
Aircraft emissions	0	0.23–0.6 (0.6)
Biomass burning	0.25–7 (1.5)	3–13 (6.0)
Lightning	3–25 (5.06)	3–25 (5.06)
Soil NO_x emissions	3.59–18.2 (4.76)	4–21 (4.76)
Natural	4–15.5	4–15.5
Agricultural	?	1.8–5.4
NH_3 oxidation	0.2–0.6	0.5–3
Stratospheric injection	0.1–0.6 (0.5)	0.1–0.6 (0.5)
Total	7.8–41 (11.82)	23–81 (36.1)

Table 2. Global emissions of NH_x in Tg N y^{-1} to the troposphere for the mid 1800s and the 1980–90s (Bottger et al. 1978; Bouwman et al. 1997; Dentener & Crutzen 1994; Galloway et al. 1995; Schlesinger & Hartley 1992; Soderlund & Svensson 1976; Stedman & Shetter 1983; Warneck 1988). Values in parentheses are those used in the simulations described below (Dentener & Crutzen 1994).

Source	Pre-industrial	Industrial
Fossil fuel combustion	0	0.1–2.2 (–)
Industrial process	0	0.2 (–)
Domestic animal excreta		20–43 (22)
Biomass burning		2.0–8.0 (2.0)
Domestic animals + biomass burning	8.95 (8.95)	
Crops	(–)	3.6 (–)
Wild animal excreta	2.5 (2.5)	0.1–6 (2.5)
Synthetic fertilizer use	0	1.2–9.0 (6.4)
Oceans		8.2–13 (8.2)
Soils and natural vegetation	3.8 (3.8)	2.4–10 (5.1)
Humans and pets		2.6–4 (–)
Total	15–21 (15.2)	45–83 (46.2)

are: the NO_2 V_d over land is 0.25 cm s^{-1} and over the sea is 0.1 cm s^{-1}; the NO V_d over land is 0.04 cm s^{-1} and over the sea is 0.0 cm s^{-1}; the HNO_3 V_d over land is 2.0 cm s^{-1} and over the sea is 0.8 cm s^{-1}; the NO_3 V_d over land is 2.0 cm s^{-1} and over the sea is 0.8 cm s^{-1} (Dentener & Crutzen 1993); sulfate aerosols including $(NH_4^+)_2SO_4^{2-}$ have a deposition velocity of 0.1 cm s^{-1}, and dry deposition of NH_3 on land surfaces was calculated in the canopy using a temperature and biomass dependent canopy compensation point (Dentener & Crutzen 1994). Wet deposition is calculated according to

$$P = \frac{\epsilon R}{L}, \tag{2}$$

where P (s^{-1}) is the wet deposition rate constant, ϵ is dimensionless parameter correcting for less soluble species (for highly soluble species, $\epsilon = 1$); L (g m^{-3}) is the liquid water content of the rain cloud; R (g m^{-3} s^{-1}) is a function of R_0, the precipitation rate at the surface. R is calculated from R_0 using the function g (m^{-1}) describing the fraction of precipitation released at a given height interval calculated from the zonal mean data on the release of latent heat (Newell et al. 1974):

$$R(\phi, \lambda, Z, t) = R_0(\phi, \lambda, Z, t) \cdot g(\lambda, Z, t), \tag{3}$$

where ϕ is latitude, λ is longitude, Z is height and t is time. The following chemical species are subject to scavenging by precipitation: HNO_3, HNO_4, CH_2O, H_2O_2, CH_3O_2H, H_2SO_4, $(NH_4)_2SO_4$, SO_2, and NH_3. The gas phase species were corrected for their solubility using Henry's law coefficient.

This description of MOGUNTIA provides a basic overview of the model. More detail on how NO_x/NO_y and NH_x chemistry is implemented in the model is available in Dentener & Crutzen 1993 and 1994. The transport scheme and the photochemical schemes are described in Zimmerman 1987 and Crutzen & Zimmerman 1991. A comparison of the deposition scheme and simulated N deposition with the schemes and results of four other 3-D chemical transport models is available in Holland et al. 1997a.

Deposition measurements

We compared the wet deposition of NH_x and NO_y simulated by MOGUNTIA to wet deposition measurements from various sources. Although comparison with dry deposition data would have been valuable, dry deposition measurements are spatially and temporally sparse and most estimates of dry deposition rely on inferential models of dry deposition (Hicks 1989; Wesely 1989). By contrast, the network data used, NADP/NTN for the U.S. and EMEP for Europe, spanned 17 years of measurements at more than

350 sites and included measurement of precipitation necessary to calculate the wet deposition flux. We also compared the MOGUNTIA results to our recent compilation of NH_x and NO_3^- wet deposition measurements over the United States and Europe (Sulzman et al. 1997; Figure 1). We compared MOGUNTIA simulated deposition with the mean of all the sites which had measured the N deposition flux within a MOGUNTIA 10° by 10° grid cell. In addition, we compared the model to a global data set of wet deposition compiled by Dentener and Crutzen (1994).

In the United States, data on deposition inputs were provided by the National Atmospheric Deposition Program/National Trends Network (NADP/NTN, see *http://nadp.sws.uiuc.edu* for more information; Figure 1). Within the U.S., precipitation was collected in buckets which were triggered to open at the onset of rain, rather than using bulk precipitation collectors, which are subject to contamination and may sample significant amounts of dry deposition. Precipitation buckets were collected weekly at between 21–203 sites throughout the United States (average 156, although between 1985 and 1994, the number of sites was close to 200). All available measurements which met the criteria of: (1) more than 90% of the data had to have simultaneous measurement of both precipitation and chemistry and (2) more than 75% valid chemical measurements, were included in the spatial analyses. The precipitation was sent to a central laboratory for chemical analyses, including hydrogen ions (acidity as pH); associated anions: sulfate, nitrate and chloride; and base cations: ammonium, calcium, magnesium, potassium and sodium.

Wet deposition data for ammonium (NH_4^+) and nitrate (NO_3^-) for Europe were provided by the Cooperative Programme for Monitoring and Evaluation of the Long-Range Transmission of Air Pollutants in Europe (European Monitoring and Evaluation Programme, EMEP; Figure 1). In Europe, precipitation was collected daily using either the precipitation only samplers similar to those described above for the U.S. or bulk sampling devices which are continually open to the atmosphere. Forty three of the 108 sites used bulk precipitation collectors. Chemical analyses were done on a site by site or country by country basis rather than at a central laboratory. Across Europe, we used data from 108 out of the 188 sites which satisfied the same criteria for completeness described above. This filtering of EMEP data was done to enhance the comparability of the U.S. and European data sets. Some of the sites including Lazaropole, formerly Yugoslavia; Neuglobsow, Germany; Jarczew, Poland; Rarau, Seminic, Paring, Fundata, Turia, and Masun, Romania; and Ivan Sedlo, Bosnia-Herzegovina collected data monthly and the data should be viewed more critically. We used all available data reported for the time period 1978–1994. Data for 1994 are published as ‘Data report 1994’ by Anne Gunn-Hjellbrekke, Jan Schaug and

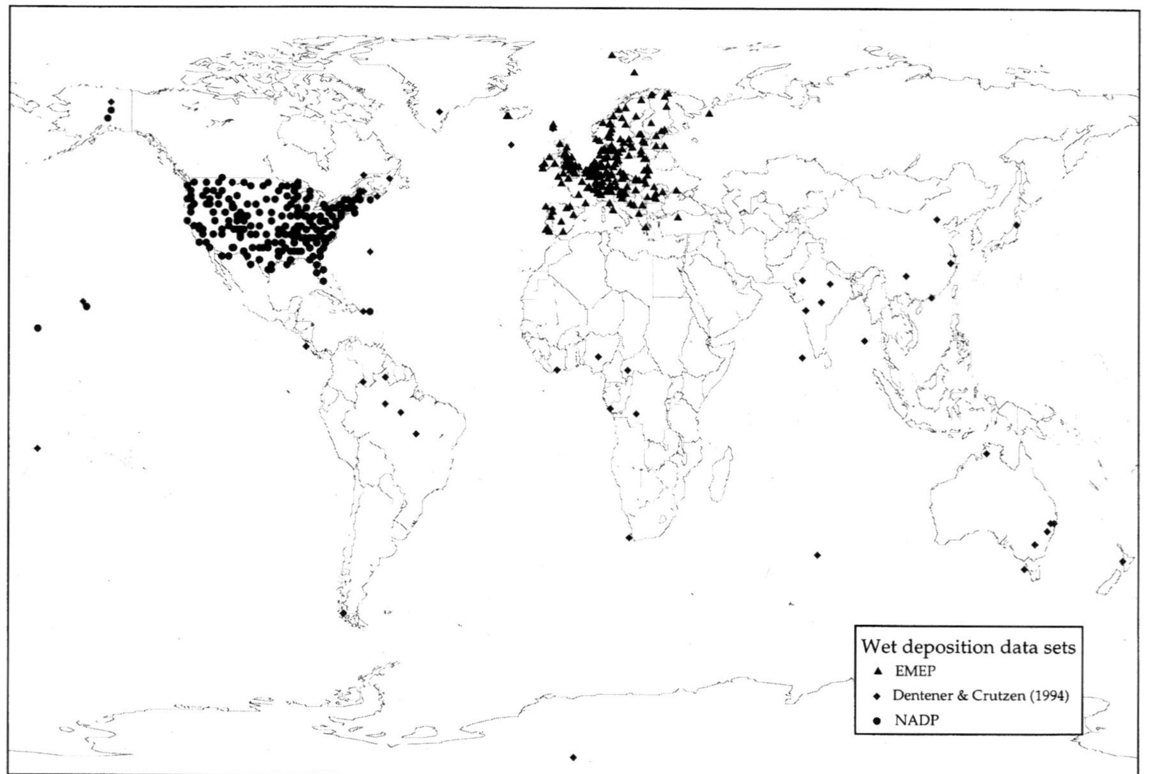

Figure 1. Map of wet deposition measurements sites used for comparison with the MOGUNTIA contemporary simulations. Three different networks were used: (1) the global network of data described in the Table 3 of Dentener and Crutzen, (2) NADP/NTN data for the U.S., and (3) EMEP network data for Europe. More details on the data and the networks is provided in the text.

Jan Erik Skjelmoen (EMEP/CCC 4/96), Kjeller, Norway 1996. Variations in collection procedures, analytical techniques, and the spatial coverage of the measurement sites as well as the representativeness of the site locations complicate the comparison across NADP and EMEP networks and amongst sites within the EMEP network. For both the European and U.S. data, we used data collected between 1978 and 1994.

To calculate the integrated deposition over the U.S. and Europe, we first mapped the wet deposition of NH_4^+ and NO_3^- by spatial interpolation of the annual mean site observations of wet deposition using moving window Kriging (Haas 1990) with precipitation and elevation covariates. The spatial interpolation (the Kriging analysis) was done using 237 sites for the wet deposition within the United States, and 108 sites for the wet deposition estimates within Europe. The spatial integration for Europe included both the bulk precipitation and wet-only deposition measurements. Some sites made measurements for only a few years, while others continued for the whole time period; thus, these numbers represent the total number of sites which made measurements in the 1978 and 1994 time period. For the U.S., we used gridded precipitation and elevation from the VEMAP Phase I data set (Kittel et al. 1995). This choice was made after evaluating the effects of using a wide range of covariates (including temperature, humidity, etc.) on Kriging model statistics. These statistics are based on a series of cross-validation studies in which each site in turn was withheld from the analysis, and the distribution of residuals (modeled minus predicted deposition) was examined. For Europe, we used precipitation and elevation from the Leemans and Cramer global data set (Leemans & Cramer 1991). All covariates and our desired output base map for both regions were defined on a $0.5° \times 0.5°$ grid. To calculate the NH_x, NO_y and $NH_x + NO_y$ wet deposition over a region we summed the wet deposition in each of the grid cells. The same summation procedure was applied to the MOGUNTIA simulated deposition, after the output was regridded to the same 0.5° by 0.5° grid.

Vegetation maps

The type of vegetation receiving the N deposition is important to determining the ecological impact of the added N (Aber & Driscoll 1997; Aber et al. 1989). We used global maps of potential and present-day vegetation to examine the quantities of N being deposited on different biomes. For examination of pre-industrial N deposition onto terrestrial ecosystems, we use the potential natural vegetation map of Cramer et al. (1995), which is the same map used for global CENTURY simulations (Schimel et al. 1996; Schimel et al. 1997), and for estimation of global N_2 fixation (Cleveland et al. submitted). For examination of contemporary N deposition to terrestrial ecosystems, we used the global map of vegetation distributions by DeFries and Townshend (1995). The vegetation classes used for the contemporary scenario included only 15 classes compared to 35 potential natural vegetation classes defined for the pre-industrial scenario.

The MOGUNTIA deposition fields and the DeFries and Townsend vegetation map were re-gridded to a 0.5° by 0.5° grid, without spatial interpolation or smoothing. For each half-degree cell a land cover type was assigned, and the deposition estimate was multiplied by the area of a vegetation type within each half-degree cell. We then calculated an areal-based summary deposition for each biome type on the common half-degree cells (Table 3).

Results and discussion

NO_x emissions

Emissions of oxidized N ($NO_x = NO + NO_2$) generate NO_3^- and NO_y deposition. The most common mechanism for NO_x removal from the troposphere is via the following reaction:

$$OH + NO_2 \longrightarrow HNO_3(g), \tag{4}$$

and via hydrolysis of N_2O_5:

$$N_2O_5 + H_2O \longrightarrow 2HNO_3, \tag{5}$$

Once formed, the HNO_3 is removed from the atmosphere via precipitation resulting in wet deposition. Gaseous HNO_3 may also be removed by dry deposition (Prather et al. 1995; Ridley & Atlas 1999).

Lightning, biomass burning and soil emissions of NO_x constitute the major sources of oxidized N in the pre-industrial atmosphere. Modeled NO_x fluxes from lightning and soil emissions are the same for the pre-industrial and contemporary time periods (Dentener & Crutzen 1994). Biomass burning NO_x fluxes are four-fold higher in contemporary simulations (Dentener & Crutzen 1994).

During the past decade, estimates of lightning production of NO_x have ranged from 2–100 Tg N y^{-1} (Table 1). Five 3-D chemical transport models that estimate deposition patterns based on emissions include global estimates of lightning production of NO_x ranges from a low of 3.0 Tg NO_x-N y^{-1} for GCTM (Levy et al. 1996a; Levy et al. 1996b) to a high of 10 Tg NO_x for GRANTOUR (Penner 1991; Penner 1994). The most recent estimates of

Table 3. Pre-industrial and contemporary N deposition onto biome types summarized from Tables 1 and 2 of the Appendix. The deposition is total deposition, $NO_y + NH_x$, and includes both wet and dry deposition simulated by MOGUNTIA (Dentener & Crutzen 1994). The land cover classification used for the pre-industrial scenario was the potential natural vegetation of Cramer et al. (1995). For the contemporary scenario, the land cover classification was the actual land cover described by Defries and Townshend (1994). The aggregation of vegetation types into biomes is indicated in the first column of Appendix 1, Tables 1 and 2. Ranges of deposition across vegetation types are provided in Appendix 1.

Biome	NH temperate latitudes			Tropical			SH temperate latitudes		
	N deposition $* 10^9$ g N	Area ha $* 10^9$	Average N deposition kg N ha^{-1}	N deposition $* 10^9$ g N	Area ha $* 10^9$	Average N deposition kg N ha^{-1}	N deposition $* 10^9$ g N	Area ha $* 10^9$	Average N deposition kg N ha^{-1}
				Pre-industrial					
Grasslands	**0.88**	*1.82*	0.50	**0.38**	*0.64*	1.42	**0.17**	*0.20*	0.63
Forests	**1.94**	*2.66*	1.02	**4.18**	*2.18*	1.85	**1.18**	*0.15*	0.76
Mixed	**0.82**	*1.70*	0.58	**2.77**	*2.09*	1.19	**0.51**	*0.80*	0.80
Life-forms wetlands & riparian zones	**0.37**	*0.66*	0.77	**0.31**	*1.48*	1.83	**0.01**	*0.02*	0.80
Ice	**0.02**	*0.05*	0.37				**0.00**	*0.00*	0.42
				Contemporary					
Grasslands	**6.39**	*2.42*	2.81	**2.87**	*1.99*	2.26	**0.32**	*0.62*	1.49
Forests	**9.47**	*2.58*	4.55	**4.94**	*1.49*	3.58	**0.19**	*0.10*	1.43
Mixed	**4.16**	*0.78*	6.20	**5.98**	*1.51*	3.94	**1.04**	*0.69*	1.85
Life-forms wetlands & riparian zones	**6.61**	*0.88*	7.42	**1.30**	*0.32*	4.15	**0.21**	*0.14*	1.54
Ice	**0.02**	*0.03*	0.55				**0.00**	*0.02*	0.05

lightning NO_x emissions of 12.2 and 13.2 Tg NO_x-N y^{-1} have been based on both lightning physics (Price et al. 1997a) and the global atmospheric electric circuit, respectively (Price et al. 1997b). Extrapolation of aircraft measurements made during thunderstorms in New Mexico, suggest global lightning emissions of 2.4–4.9 Tg NO_x-N y^{-1} between 8 and 12 km in altitude, but do not include the NO_x production rate below that altitude (Ridley et al. 1996). IPCC 1994 estimated lightning production of NO_x to be 8 Tg NO_x-N y^{-1} (Prather et al. 1995), while the 1994 Scientific Assessment of Ozone Depletion estimated lightning NO_x fluxes to be 7 Tg NO_x-N y^{-1} with a range of 3–20 Tg NO_x-N y^{-1} (Sanhueza et al. 1995), underscoring the uncertainty. There is substantial agreement that lightning emissions of NO_x lie somewhere between 3 and 25 Tg NO_x-N y^{-1}, with the most likely estimates falling between 10 and 15 Tg NO_x-N y^{-1}. This is considerable progress from a few years ago, when the estimates reached 100 Tg NO_x-N y^{-1} (Franzblau & Popp 1989).

Estimates of contemporary soil NO_x emissions range from 4–21 Tg NO_x-N y^{-1} (Table 1). Soil NO_x emission estimates used in five different global chemical transport models range from 4–10 Tg NO_x-N y^{-1} (Holland et al. 1997b), and some of the variability amongst the estimates can be explained by whether a soil NO_x-N emission estimate considers canopy uptake of NO_x. Both the ECHAM model estimate of 10 Tg NO_x-N y^{-1} and the MOGUNTIA estimate of 4.76 are based on the same studies (Galbally & Johansson 1989; Galbally & Roy 1978), but ECHAM neglects canopy scavenging of NO_x. Two models of soil NO_x-N emissions estimate the global soil NO_x flux to be 9.7 and 10.2 Tg NO_x-N y^{-1} (Potter et al. 1996; Yienger & Levy II 1995 respectively). The Yienger and Levy estimate was reduced to 5.45 Tg NO_x-N y^{-1} when parameterization of canopy scavenging was included in the simulation. However, a recent compilation of available measurements of soil NO_x-N emissions estimated below-canopy emissions of NO_x-N to be 21.2 Tg NO_x-N y^{-1} and above-canopy emissions of NO_x-N to be 13 Tg NO_x-N y^{-1} (Davidson & Kingerlee 1997). The likely error of the measurement compilation estimate is between 4 and 10 Tg NO_x-N y^{-1}. The error is attributable to uncertainties in estimated land areas, particularly for tropical woodlands and savannas, as well as a limited number of measurements in some biomes and vegetation types, particularly tropical agricultural ecosystems and deserts.

Production of NO_x-N related to biomass burning may be divided into two parts: that NO_x-N emitted during burning, and the NO_x-N emitted from the soil post-burning. Both fire-related sources of NO_x have increased with human activity. The emission of NO_x following biomass burning is neglected in most of the global compilations (both the soil and biomass burning portions of the budget), because clear documentation of the increase in flux and its duration were measured only recently (Neff et al. 1995; Veldkamp & Keller 1997). The major sources of uncertainty associated with estimation of NO_x-N production during biomass burning are the quantification of the area burned, biomass burned, the timing and duration of the burn, and the amount of NO_x produced per unit of CO_2 released during the burn (Table 1). Crutzen and Andreae (1990) estimate 2.1 and 5.5 Tg NO_x-N y^{-1} emitted by biomass burning globally, with the greatest contribution by tropical fires. In 1987, burning in the Amazon alone contributed 1 Tg NO_x N y^{-1} (Setzer & Pereira 1991). A current inventory of biomass burning fluxes of both NO_x and NH_3 is underway as part of the Global Emissions Inventory Activity (GEIA, *http://blueskies.sprl.umich.edu/geia/*). Estimates of modern day contributions of biomass burning to the global NO_x-N budget differ by 4-fold (Table 1) and the pre-industrial estimates of biomass burning contributions range from 0.25 to 7 Tg NO_x-N y^{-1}, an even greater range. The difficulty of quantifying the area and biomass burned is a problem for both the pre-industrial and contemporary estimates of NO_x and NH_3 emissions.

Combustion of fossil fuel and aircraft NO_x emissions are by-products of industrialization, and have increased from almost 0 in the last century to between 20 and 25 Tg N for the last two decades. Ninety percent of industrial NO_x emissions originate from countries located at northern temperate latitudes (the U.S., Canada, Western and Eastern Europe, the Former USSR, China, Japan, and the Middle East) (Olivier et al. 1995). The energy statistics and emission factors (NO_x per unit CO_2 emitted) used to quantify fossil fuel combustion for both NO_x and CO_2 are known, but the emission factors vary by an order of magnitude from fuel type to fuel type. Uncertainty arises from limitations of reporting fossil fuel consumption statistics on a country by country basis, and the re-gridding necessary for use in 3-D global models. The differing years for which statistics are reported also contribute to the uncertainty. There have been a number of global compilations of fossil fuel NO_x emissions (Benkovitz et al. 1996; Dignon & Hameed 1989; Logan 1983; Muller 1992). The Benkovitz et al. (1996) compilation is available through GEIA (*http://blueskies.sprl.umich.edu/geia/*). Uncertainties for fossil fuel NO_x emissions remain ±30%, and may be as high as 50% for some regions of the world (Lee et al. 1997). Uncertainties for the potentially large natural sources (e.g. biomass burning, lightning and soil emissions), remain at ±100% and greater as shown here (Table 1).

NH_3 emissions

NH_x exchanges are important because of their magnitude (Table 2). NH_3 is involved in aerosol formation, plays a central role in the global nitrogen cycle, and is the most abundant atmospheric base with the ability to neutralize harmful acids. In addition, when NH_3 or NH_4^+ is oxidized in soils, through nitrification, acidifying hydrogen ions are liberated (van Breemen et al. 1982). Estimates of global NH_3 emissions exceed those for NO_x (Bottger et al. 1978; Bouwman et al. 1997; Dentener & Crutzen 1994; Galloway et al. 1995; Schlesinger & Hartley 1992; Soderlund & Svensson 1976; Stedman & Shetter 1983; Warneck 1988). Fossil fuel contributions to global NH_3 emissions are small, but domestic animal excreta dominates both the pre-industrial and contemporary NH_3 budgets (Table 2). Present day domestic animal populations are quantified to within 10%, but pre-industrial estimates rely on extrapolations through time based on a much sparser data set or by analogy to the better quantified human population (Bouwman et al. 1997; Olivier et al. 1995). The total estimate of uncertainty for NH_3 emissions from domestic animal excrement is ±50%, which is reflected in the range of the estimates in Table 2. Emission of NH_3 from well fertilized crops and from soils following the application of animal waste and synthetic fertilizer is substantial. The percentage of fertilizer returned to the atmosphere as NH_3 varies between 2 and 30% depending on the type of fertilizer, soil characteristics, and fertilizer management, particularly the timing method of application. Although, fertilizer use is well quantified (FAO 1985; Matthews 1994), estimates of NH_3 emissions following fertilization vary (Bouwman et al. 1997; Warneck 1988). Furthermore, the most recent estimates of 5–9 Tg N y^{-1} are much greater than the previous estimates of 1.2–3.5 Tg N y^{-1} (Bottger et al. 1978; Crutzen & Gidel 1983; Stedman & Shetter 1983; Warneck 1988), and the differences cannot be explained by the rise in fertilizer use between the two sets of estimates. Crop emissions are estimated to be 2.5 kg N ha^{-1} for all arable land. The extensive data available for NH_3 emissions from a wide array of industrial processes including fertilizer and chemical manufacture allow for the robust determination of a rather small number (Bouwman et al. 1997). Estimation of NH_3 emission by fossil fuel combustion is limited by the availability of algorithms for different fuel types, but the most recent estimate of Bouwman et al. (1997) is quite low at 0.1 Tg N y^{-1} with an uncertainty of 0.0–0.3 Tg N y^{-1}. All animals, including humans and their pets, emit NH_3 in their breath and sweat. The latest global estimate of 2.6 Tg N y^{-1} (Bouwman et al. 1997) assumed an emission of 0.5 kg N $person^{-1}$ (or animal) y^{-1}. Agriculture contributes more than 50% of the NH_3 emissions globally. Emissions of NH_3-N from sub-tropical and tropical latitudes are much more important than they are for global NO_x-N, which are dominated by Northern Hemisphere fossil fuel emissions.

As with the NO_x emissions budgets, quantification of the natural sources of NH_3, including emissions of NH_3 from the excreta of wild animals, soils and natural vegetation, and the ocean, are uncertain, and estimates vary by two to four-fold. These natural sources comprise the pre-industrial NH_3 budget and so their uncertainty dominates estimation of pre-industrial NH_3 emissions. Many factors contribute to the uncertainty. Wild animal population censuses are rare and global censuses even rarer (Bouwman et al. 1997). Soil and vegetation NH_3 emission estimates are complicated by plant scavenging of atmospheric NH_3. Over the course of a few days, plant uptake can completely counterbalance emissions (Langford et al. 1992). Global soils emissions have been estimated as high as 38 Tg N y^{-1} (Dawson 1977) but it is unlikely that all of the NH_3 emitted escapes to the atmosphere (Warneck 1988). Oceanic emissions estimates suffer the same problem as plant and soils emission because NH_3 may be either consumed or emitted (Bouwman et al. 1997). Thus, the sole estimate of pre-industrial NH_3 emissions made by Dentener and Crutzen (1994) is highly uncertain. A complication is the extent to which natural emissions and associated deposition of NO_x and NH_3 from soils/vegetation and biomass burning represent a net input into an ecosystem.

Modeled nitrogen deposition

Simulated pre-industrial N deposition was greatest over tropical ecosystems according to the MOGUNTIA simulations (Figure 2; Dentener & Crutzen 1994). Rates of total N (NH_x + NO_y) deposition ranged from 0–3.1 kg N ha^{-1} y^{-1} before industrialization with the highest rates of deposition in northwestern South America, Central Africa, and Southern Asia including the Indonesian Islands. Total NH_x deposition was highest over northeastern South America, and greater than 0.5 kg ha^{-1} over large areas of tropical Africa, South America, and Asia, and less than 1 kg N ha^{-1} for both northern and southern temperate ecosystems. Total NO_y deposition was greatest in Central Africa, greater than 0.75 kg N ha^{-1} over most of the tropical land mass, and less than 1 kg N ha^{-1} for both northern and southern temperate ecosystems. These high rates of pre-industrial N deposition onto tropical latitudes are driven by biomass burning and soil emissions of NO_x and NH_3 as well as lightning production of NO_x, all of which are greatest in the tropics (Holland 1997a). Thus, the pre-industrial pattern of high emissions and deposition at tropical latitudes causes a continental scale recycling of N in ecosystems where the only other source of N is via biological N_2 fixation. However, the rates of N fixation for tropical ecosystems are estimated to be between 10 and 20 kg N ha^{-1} y^{-1} (Cleveland et al. submitted), 10 times the average pre-industrial N deposition.

Pre-industrial and contemporary total N deposition had strikingly different magnitudes and spatial patterns (Figures 2 and 3). In contrast with the tropical distribution of the greatest pre-industrial total N deposition, contemporary total N deposition is greatest at northern temperate latitudes (Figures 2 and

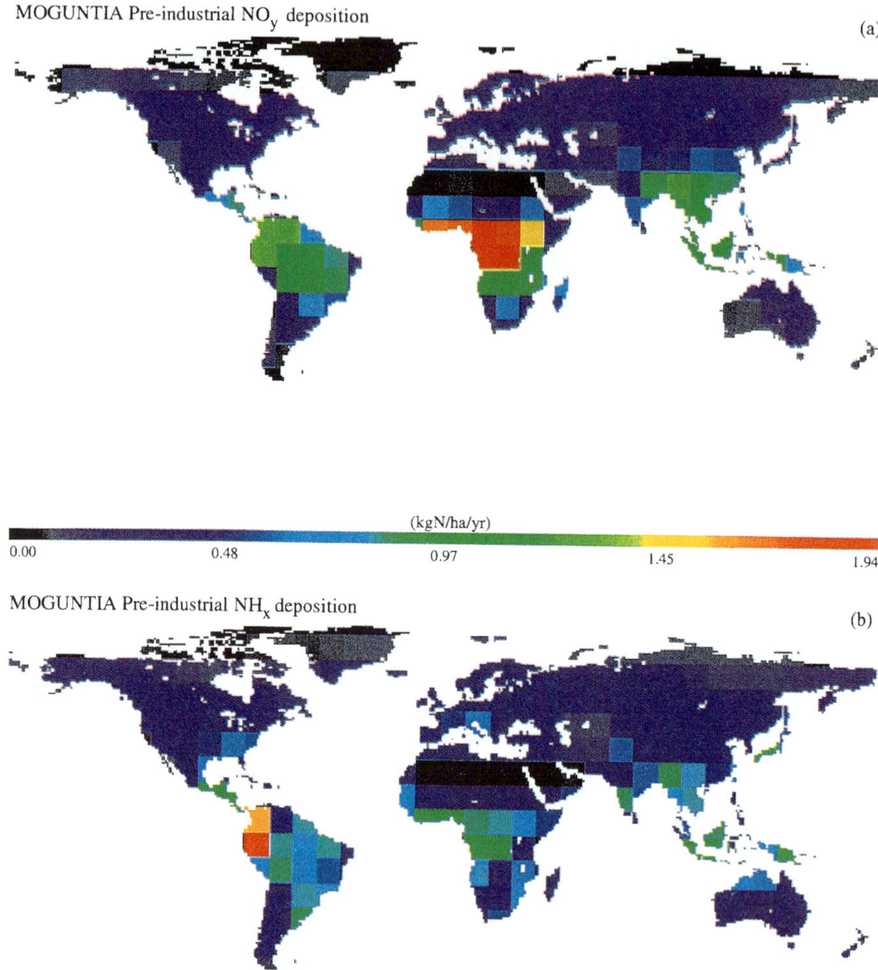

Figure 2. Modeled pre-industrial wet and dry NO_y-N deposition (a) and NH_x-N deposition (b) deposition onto terrestrial ecosystems. The results were generated using a model of the troposphere, MOGUNTIA (Dentener & Crutzen 1994).

3; Table 3; Appendix 1; Dentener & Crutzen 1994). Across all of the biomes at tropical and temperate latitudes, contemporary total N deposition exceeds pre-industrial total N deposition (Table 3). The northern temperate biomes most affected by total N deposition, cultivated lands and mixed forests, receive >16 times more N today than they did a century ago according to the MOGUNTIA simulations. On average, there was more than a four-fold increase in the rates of N deposition onto NH temperate ecosystems for the contemporary compared to the pre-industrial scenario. The increase in nitrate concentration in NH ice cores shows a similar increase over the last century

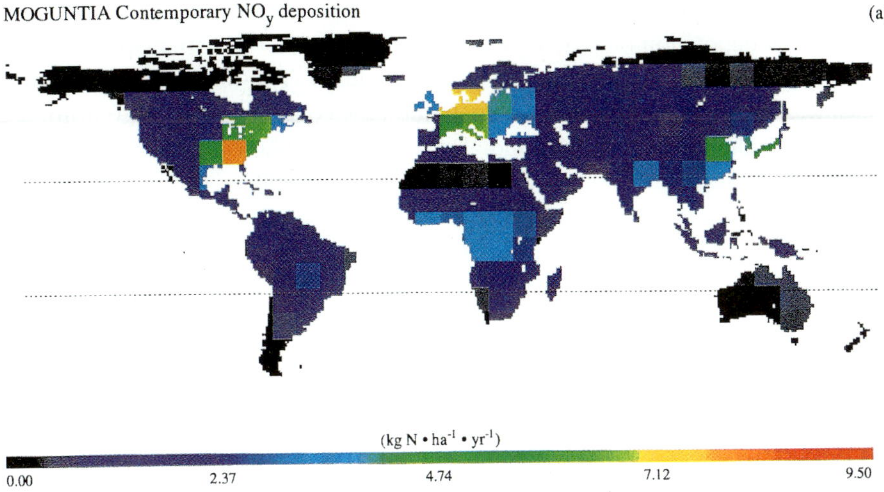

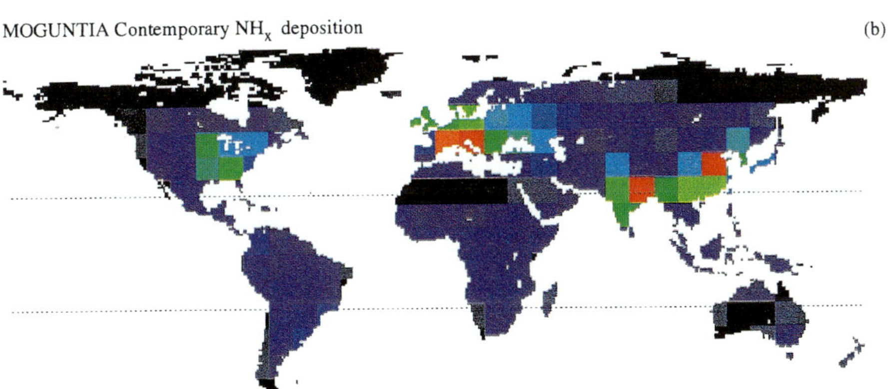

Figure 3. Modeled contemporary wet and dry NO_y-N deposition (a) and NH_x-N deposition (b) deposition onto terrestrial ecosystems. The results were generated using a model of the troposphere, MOGUNTIA (Dentener & Crutzen 1994).

(Mayewski et al. 1990; Wagenbach et al. 1988). Total NO_y deposition exceeds 10 kg N ha^{-1} y^{-1} over the Eastern U.S. and Western Europe, which are among the most industrialized regions of the modern world. Southern Asia, one of the fastest developing regions of the world, is receiving more than 5 kg N ha^{-1} annually. By contrast, the increase in the average rate of N deposition onto tropical and SH temperate ecosystems was more modest for the contemporary compared to the pre-industrial scenario. The MOGUNTIA simulations demonstrate a tremendous increase in global total N deposition over the last century and a half corresponding to the expansion of industrialization, per capita consumption of N, agriculture, and the world population.

For both the pre-industrial and contemporary N deposition scenarios, the range of N deposition onto an ecosystem or vegetation type often spanned an order of magnitude (Appendix 1, Tables 1 and 2). The range in deposition rates was much narrower when a vegetation type covered less than 20 million hectares within a latitudinal band. Because the lifetime of many reactive nitrogen species is relatively short, hours to days for NH_3, 4.5 days for NH_x (Dentener & Crutzen 1994), and one day for NO_x (Prather et al. 1995), a large fraction of the N tends to be deposited near the location where it was emitted. Some grid cells containing a specific vegetation type were located near modeled sources while others were remote. The distribution of sources and atmospheric dynamics thus determine the amount of N an ecosystem receives through deposition. The vegetation type itself influences the amount of N scavenged from the atmosphere via deposition velocity (see Equation 1). Vegetation may also influence deposition via emission. Some vegetation types, like savannas, emit a great deal of NO_x via soil emission and biomass burning, and the NO_x generated is mostly deposited back onto a savanna, since the lifetime of NO_x and its reaction products are rather short. Intensity of emission, and vegetation interact with the short lifetime of these gases to generate the within biome variability demonstrated by this coarse resolution model. The simulated average rates of deposition onto the biomes should be considered against this background of high spatial variability.

Model/measurement comparison

Globally, modeled and measured rates of N wet deposition were often different. Modeled NO_3^- deposition agreed well with measurements in the United States (Figure 5a). By contrast, the model predicted fluxes of 50% of the measured for NO_3^- fluxes for the European wet deposition data (EMEP) and the compilation of global measurements made by Dentener and Crutzen (1994) (Figures 6 and 7a). Modeled NH_x deposition followed a similar pattern. For the U.S. network, the model predicted wet deposition fluxes of NH_4^+ that were 97% of the measured wet deposition (Figure 4b). However, the model predicted only 43% of the measured deposition for the European network, and only 37% of the measured NH_x deposition for the Dentener and Crutzen (1994) global compilation (Figures 6b and 7b). In addition, the low r^2 for the European (EMEP) NH_x and NO_y comparisons suggests that the model was also unable to capture the correct spatial pattern of wet deposition. These results suggest that model tends to underestimate wet deposition over Europe and in remote areas, but agrees well with wet deposition measurements in the U.S.

Incorrect partitioning of wet and dry deposition may explain some of the discrepancies between the model and measurement data. In Europe, modeled

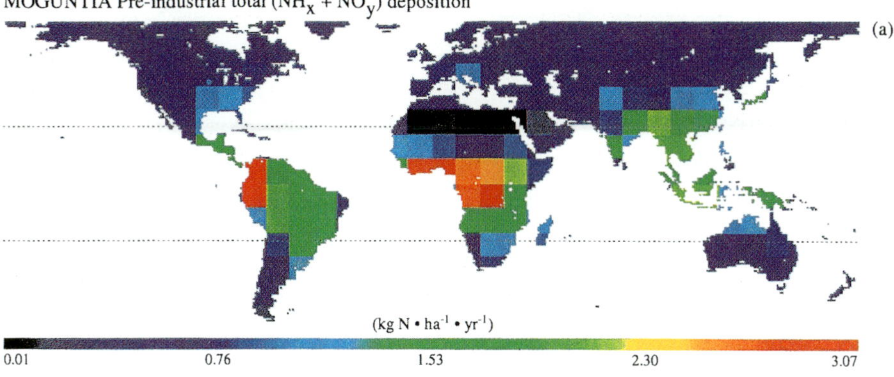

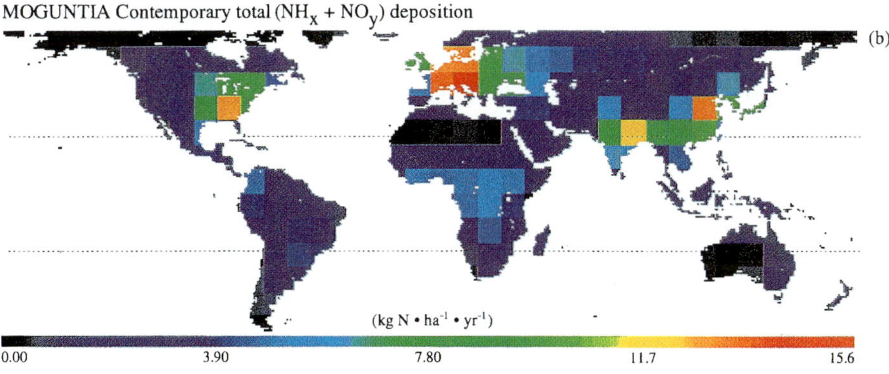

Figure 4. Modeled pre-industrial (a) and contemporary (b) total global N deposition including wet and dry deposition of both NO_y-N and NH_x-N. The results were generated using a model of the troposphere, MOGUNTIA (Dentener & Crutzen 1994).

total deposition of NH_x and NO_y corresponded to bulk precipitation measured deposition (Figure 6a and b). However, bulk deposition measurements of NH_x and NO_3^- are only 4–34% higher than wet-only deposition measurements, because they do not fully capture dry deposition inputs. Thus, differences in methodology offer only a partial explanation of model/measurement discrepancies. To further examine the partitioning of wet and dry deposition, we compared the spatially interpolated wet deposition over Europe and the U.S. with estimated emissions, and the wet and dry deposition fluxes predicted by this coarse resolution model (Table 4). Over both the United States and Europe, dry deposition of NH_x was 50% of total deposition, and dry deposition of NO_y was 65 and 59% of total deposition for the U.S. and Europe, respectively. By comparison, a recent WMO compilation of global deposition measurements found dry deposition of oxidized N (HNO_3 and particulate

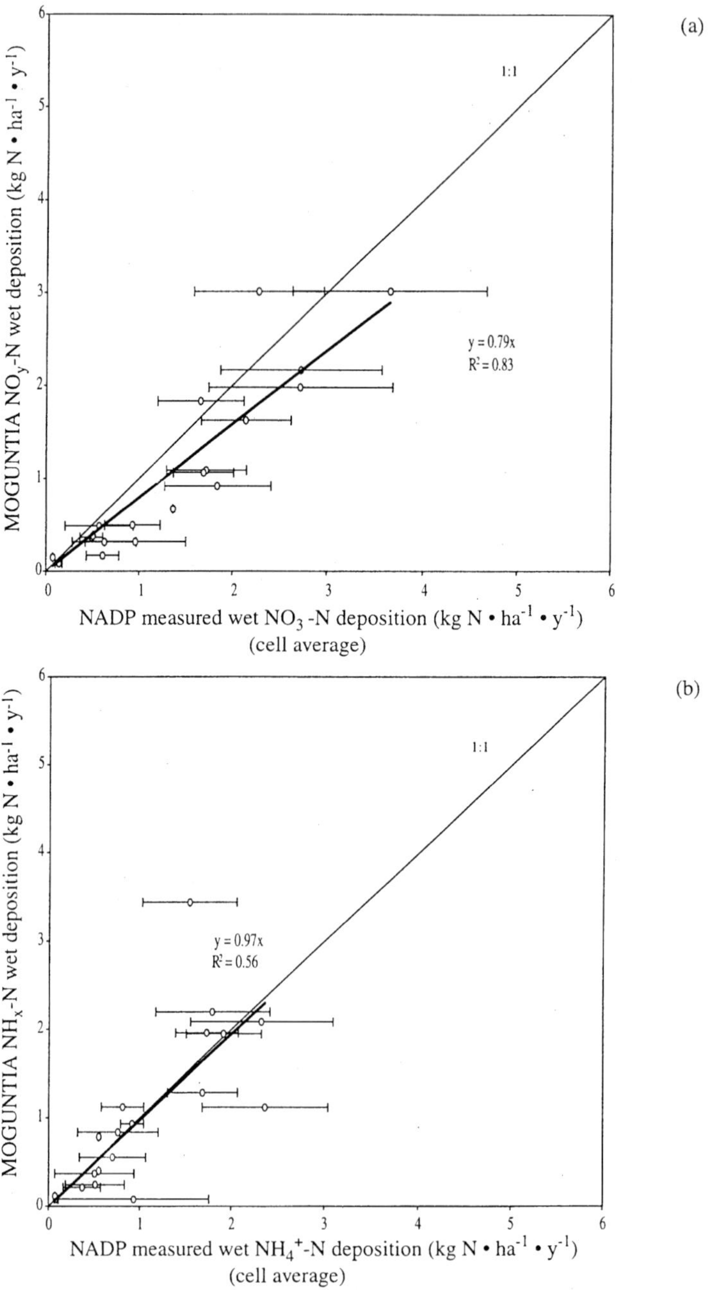

Figure 5. Comparison of simulated NO_y deposition in precipitation with measured wet NO_3^- deposition (a) and of simulated NH_x deposition in precipitation with measured wet NH_4^+ deposition (b) for the United States (NADP/NTN). The bars represent +/– standard deviation of the measured annual N deposition for the sites within a single grid cell. The standard deviation reflects only the spatial variation.

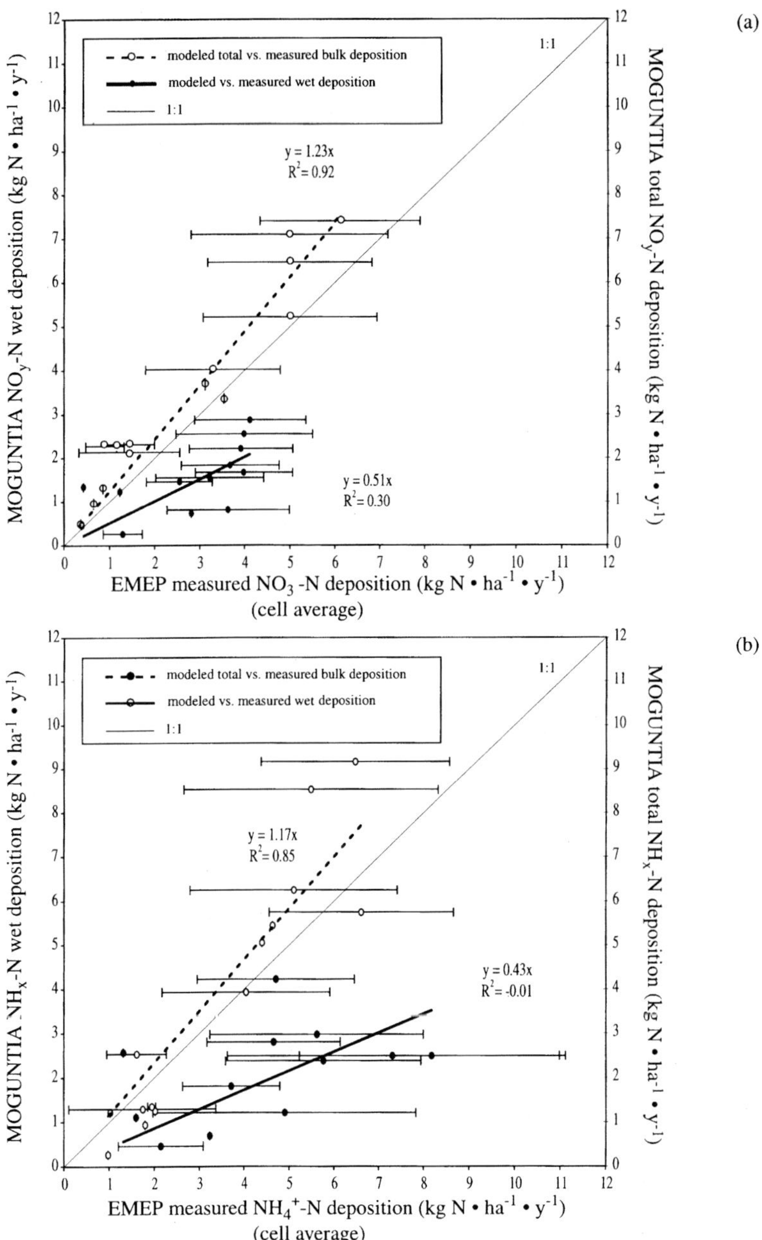

Figure 6. Comparison of European (EMEP) measurements with simulated deposition. (a) Comparison of simulated NO_y wet deposition with measured wet NO_3^- deposition and simulated NO_y wet + dry deposition with measured bulk NO_3^- deposition rates, and (b) Comparison of simulated NH_x wet deposition with measured wet NH_4^+ deposition and simulated NH_x wet + dry deposition with measured bulk NH_4^+ deposition rates. The bars represent +/– standard deviation of the measured annual N deposition for all of the sites within a single grid cell. The standard deviation reflects only the spatial variation.

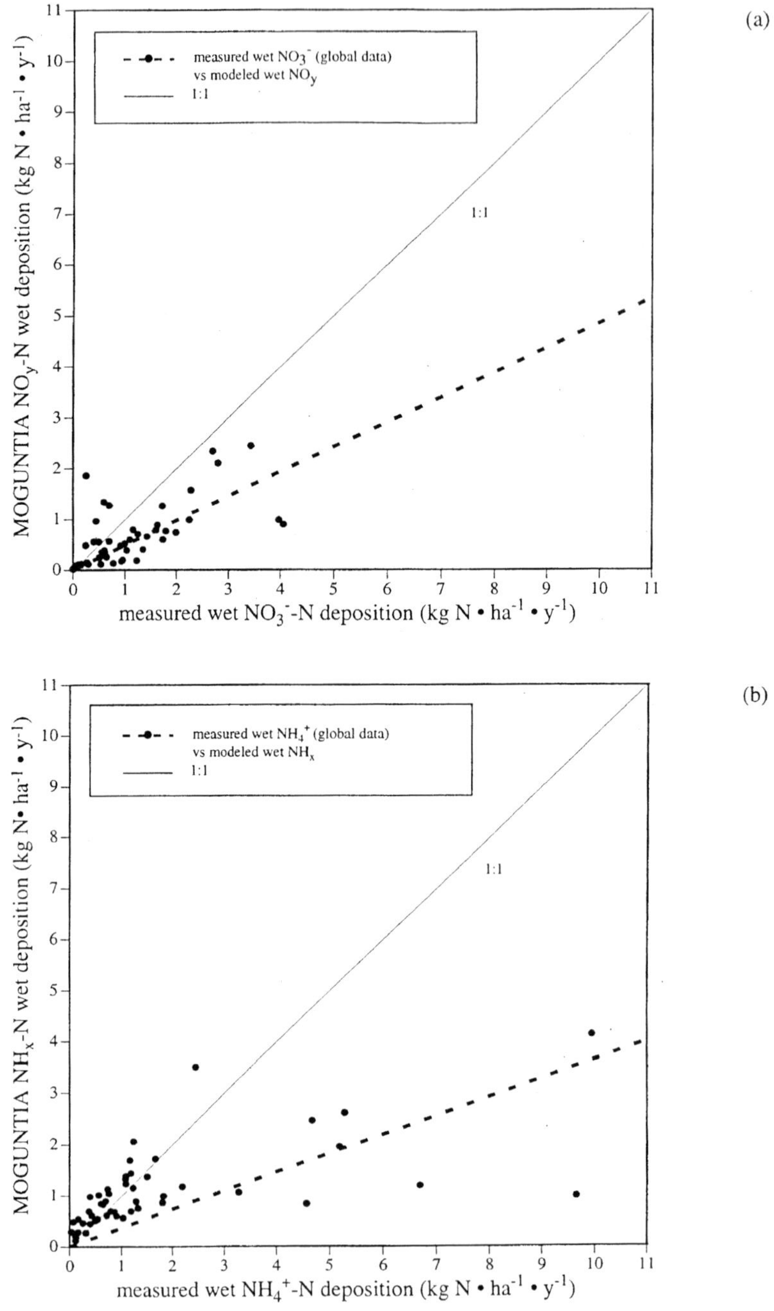

Figure 7. Comparison of simulated NO_y wet deposition, wet + dry NO_y deposition and measured wet NO_3^- deposition (a), and of simulated NH_x wet deposition, wet + dry NH_x deposition total with measured wet NH_4^+ deposition (b) for the compilation of global measurements (Dentener & Crutzen 1994).

Table 4. Spatially integrated emissions, measured wet deposition from NADP/NTN and EMEP, and modeled wet, dry and total deposition. More information on the measurements and integration is provided in the methods section.

	NO_x/NO_y Tg N y^{-1}	NH_3/NH_x Tg N y^{-1}	NO_x/NO_y + NH_3/NH_x Tg N y^{-1}
	United States[1]		
Emissions[2]	6.3	5.2	11.5
Measured wet	1.42	1.18	2.60
Deposition modeled:			
wet deposition	0.94	1.04	1.98
dry deposition	1.72	1.03	2.75
total deposition	2.66	2.07	4.73
	Europe[3]		
Emissions[2]	6.07	5.2	11.27
Measured wet	2.34	3.96	6.30
Deposition modeled:			
wet deposition	1.26	1.81	3.07
dry deposition	1.83	1.85	3.69
total deposition	3.09	3.66	6.76

[1] Land area of the U.S. considered is 7.77 million km^{-2}.
[2] Fossil fuel NO_x emissions are according to Benkovitz et al. 1996 from the NAPAP and EMEP inventories. NH_3 emissions for the U.S. are for all of North and 'Middle' America according to Dentener & Crutzen 1994.
[3] Land area of Europe considered is 8.99 million km^{-2}.

NO_3^-) to be 30–50% of wet + dry oxidized N deposition over the United States, and 20–50% of wet + dry oxidized N deposition over Europe (Whelpdale et al. 1996; Whelpdale et al. 1997). The MOGUNTIA partitioning of wet and dry deposition was more biased toward dry deposition, and likely contributed to the model/measurement data discrepancy for NO_y.

The partitioning hypothesis is consistent with the fact that these MOGUNTIA simulations did not consider deposition of particulate nitrates. There is a growing body of evidence that particulate nitrates (e.g. ammonium nitrate) are present in significant amounts (Erisman & Draaijers 1995). The deposition velocity of particulate nitrate is lower by a factor of 5–10; and hence, wet deposition may be a more important removal mechanism. Underestimation of wet NH_x deposition may be explained by problems in correctly representing the oxidized sulfur cycle in MOGUNTIA, which is usually referred to as the 'oxidant limitation' problem. An underestimate of NH_3

uptake on acidic sulfuric estimate may lead to an increased gas phase NH_3 deposition, and incorrect partitioning of dry and wet deposition fluxes.

Another constraint on our understanding is provided by a look at the total emission and deposition (wet + dry) budgets for the two regions where we have measurements: the U.S. and Europe. Interestingly, estimated NH_3 and NO_x emissions for the U.S. are more than 2 times the total of interpolated wet N deposition for both NH_x and NO_y deposition (Table 4). The Whelpdale et al. 1996 and 1997 compilations suggest that wet deposition in the U.S. (and Europe) is between 50 and 70% of total deposition for oxidized N. Thus, increasing integrated wet deposition by the fraction of dry deposition provides a simple, albeit uncertain, way of estimating total N deposition. Accordingly, the wet + dry oxidized N deposited onto the United States is likely to be between 2.03 and 2.84 Tg NO_y-N y^{-1}. For the reduced species, the picture is somewhat more complicated because the mechanism for atmospheric removal depends on the emissions and the chemical environment (Asman 1994). In areas of high emissions, the dominant removal is via dry deposition, and in areas of low emissions, the dominant removal is via wet deposition (Asman 1994). A single measurement suggests that wet deposition of NH_3 is 70% of total (Harrison & Allen 1991) and model calculations suggest that wet deposition is between 56 and 86% of total deposition (Asman & Van Jaarsveld 1992). Application of these multiplication factors (1/0.56 and 1/0.86) estimate wet + dry reduced N deposition onto the U.S. to be between 1.37 and 2.11 Tg NH_x-N y^{-1}. Thus, we estimate total wet + dry N deposition onto the United States, including both oxidized and reduced N species, to be between 3.40 and 4.95 Tg N y^{-1}.

In the U.S., spatially interpolated total (wet + dry) deposition measurements are only 30–43% of the estimated emitted N, and modeled total deposition onto the U.S. is only 53% of the emitted N. Simulated deposition is sensitive to emission estimates (Dentener & Crutzen 1993; Dentener & Crutzen 1994; Holland et al. 1997b; Holland & Lamarque 1997a). As pointed out above, the emission estimates themselves are uncertain. For example, Bouwman et al. 1997 estimates U.S. NH_3 emissions to be 3.6 Tg NO_x-N y^{-1} compared to the 5.2 Tg NH_x-N y^{-1} used in these simulations. The imbalance between estimated N emissions and deposition measurements may be explained by off-shore deposition of N. Using a simple model, Whelpdale and Galloway (1994) argue that the U.S. exports 0.8–1.2 Tg of oxidized N onto the North Atlantic. A more recent of estimate of the deposition onto the North Atlantic Ocean (excluding the continental shelf) is higher at 4.3 Tg N with deposition onto the Caribbean estimated to be another 4.3 Tg N (Prospero et al. 1996). Inclusion of offshore export places total N deposition within the uncertainty of the estimated emissions. An additional explanation may be that the measurement sites are located far from urban centers and thus do not reflect the substantial impact that cities and metro-agro-plexes have on regional N deposition patterns (Chameides et al. 1994). The most likely scenario is that off-shore transport, uncertainties in emissions, and the representativeness of the sites together contribute to the calculated imbalance in the U.S. N deposition/emission budget.

Estimated NH_3 emissions for Western Europe are 1.24 Tg N greater than the interpolated NH_x wet deposition measurements. Consistent with the grid cell comparisons of Europe, modeled NH_x wet deposition is less than half of the interpolated NX_x wet deposition (Table 4). Using the same procedure described for estimation of US total NH_x deposition, wet + dry NH_x deposition onto Europe is between 4.60 and 7.07 Tg NH_x-N y^{-1}. Estimated NH_3 emissions balances estimates of wet + dry NH_x deposition within the uncertainty of each. Emissions of oxidized N exceed wet + dry oxidized N deposition estimates of 3.34–4.68 Tg NO-N y^{-1} (calculating using the procedure outlined for the U.S. The inclusion of the bulk precipitation measurements may also contribute to balancing the emissions and deposition budgets. Moreover, European wet deposition measurement sites may capture some of the urban influence missed in the U.S. because the population density is so much greater in Europe. In Europe, oxidized and reduced N emissions come much closer to balancing N deposition given the substantial uncertainties, and suggests that Europe exports much less of its N than does the U.S.

Conclusions

The magnitude and spatial distribution of N deposition has changed substantially over the last 150 years. The greatest rates of pre-industrial N deposition were in the tropics, while contemporary rates of N deposition are highest in NH temperate ecosystems. Some NH temperate ecosystems, cultivated lands and mixed forest, receive 16 times more N now than they did before industrialization, but the average increase in N deposition over NH temperate ecosystems was four-fold. The fate of deposited N varies with ecosystem type and degree of N loading. Contemporary NH temperate forests are able to retain between 20 and 100% of the deposited N depending on the history of N deposition, land use history, and soil texture (Aber & Driscoll 1997; Gosz & Murdoch 1998; Nadelhoffer et al. in press; Nadelhoffer et al. 1995). The NH grasslands and tundra are able to retain between 50 and 100% of the deposited N, an ever greater proportion than forests (Gosz & Murdoch 1998). The ability of cultivated lands to retain nitrogen is limited because the only storage reservoir for nitrogen is soil organic matter which often declines with cultivation. Furthermore, agricultural lands are fertilized with N and may not be able to retain the additional deposited N because these ecosystems are already saturated with N (Holland 1997b; Townsend 1996). On cultivated lands, the remainder of the nitrogen is removed through harvest, returned to the atmosphere through trace gas production, or lost to groundwater, streams, rivers, and oceans through runoff and leaching. The observation that large quantities of N are deposited onto a relatively small area of cultivated lands may partially explain the strong correlation between NO_y deposition and riverine N fluxes (Howarth et al. 1996; Howarth, this issue).

Our comparisons of measured and modeled deposition suggest that global N deposition modeled by the Dentener and Crutzen 3-D chemical transport model (1993 and 1994) may under-estimate deposition in some regions when compared to measurements. Furthermore, examination of the U.S. and European N emission/deposition budgets point out important imbalances. In the U.S., estimated emissions exceed interpolated total deposition by 3–6 Tg N, suggesting that substantial N is transported offshore and/or the remote and rural location of the sites may fail to capture the deposition of urban emissions. In Europe, interpolated total N deposition came much closer to balancing emissions. Comparison of regional and global modeled deposition with the available measurements will continue to provide critical tests of our understanding of global and regional N cycles.

Acknowledgements

The authors are grateful to Elizabeth Sulzman, A. R. Townsend and two anonymous reviewers for their thoughtful and helpful comments. R. Staufer assisted with figure and manuscript preparation. E. A. Holland is also grateful to A. R., K. E. and D. S. Schimel who assisted with the completion of the manuscript. The work was funded by the Methods and Models for Integrated Assessment Program (MMIA) of the National Science Foundation, grant # ATM 9793346, the National Center for Atmospheric Research, and the Max-Planck-Institute für Biogeochemie. The National Center for Atmospheric Research is operated by the University Corporation for Atmospheric Research under the sponsorship of the National Science Foundation.

Appendix 1, Table 1. Pre-industrial N deposition ($NH_x + NO_y$) onto potential natural vegetation (Cramer 1995) for Northern Hemisphere (NH) temperate latitudes, tropical latitudes and Southern Hemisphere (SH) temperate latitudes. Deposition was simulated by MOGUNTIA (Dentener and Crutzen 1994). The surface emissions of NH_3 and NO_x used in the simulations are described in Tables 1 and 2. Nitrogen deposition onto individual biomes is summarized in Table 3. The biome classifications are grasslands – G, forests – F, wetlands and riparian zones – W, and mixed life forms – M. Zonal summaries of NO_y, NH_x and $NO_y + NH_x$ deposition are also included. The numbering of the vegetation types follows the numbers assigned by Defries and Townshend 1994.

Vegetation type (biome classification)	NH temperate latitudes			Tropical latitudes			SH temperate latitudes		
	Average deposition kg N ha^{-1} (range)	N deposition Gg (10^9)	Area $ha^{-1} * 10^6$	Average deposition kg N ha^{-1} (range)	N deposition Gg (10^9)	Area $ha^{-1} * 10^6$	Average deposition kg N ha^{-1} (range)	N deposition Gg (10^9)	Area $ha^{-1} * 10^6$
1. Ice	0.37 (0.22–1.12)	**18.27**	*47.22*				0.42 (0.17–0.58)	**0.84**	*1.99*
2. Polar Desert/ Alpine Tundra (G)	0.60 (0.21–2.13)	**214.4**	*311.94*	1.39 (0.84–3.08)	**45.70**	*32.62*	0.68 (0.16–0.84)	**31.75**	*45.7*
3. Wet/Moist Tundra (G)	0.39 (0.21–2.13)	**123.69**	*262.79*	2.93 (2.93–2.93)	**0.90**	*0.31*	0.17 (0.11–0.24)	**0.77**	*4.49*
4. Boreal Forest (F)	0.50 (0.24–2.13)	**637.44**	*1234.9*				0.34 (0.16–0.58)	**1.26**	*3.61*
5. Forested Boreal Wetland (W)	0.50 (0.32–0.73)	**4.48**	*8.25*				0.24 (0.16–0.58)	**3.54**	*14.48*
6. Boreal Woodland (M)	0.38 (0.21–0.84)	**198.67**	*519.87*	1.55 (0.84–3.08)	**36.58**	*23.54*			
7. Non-forested Boreal Woodland (M)	0.46 (0.21–0.84)	**28.79**	*62.40*						

8. Temperate Mixed Forest (F)	0.81 (0.16–2.13)	**427.33**	*520.26*	2.46 (0.84–3.08)	**19.61**	*7.92*	0.84 (0.84–0.84)	**1.64**	*1.96*
9. Temperate Coniferous Forest (F)	0.69 (0.16–2.13)	**165.87**	*227.73*	1.14 (0.56–1.90)	**25.89**	*22.71*	0.49 (0.44–0.58)	**0.35**	*0.72*
10. Temperate Deciduous Forest (F)	0.74 (0.39–1.90)	**266.49**	*351.15*	1.96 (1.06–3.08)	**20.86**	*10.60*	0.46 (0.17–0.58)	**3.32**	*7.24*
11. Temperate Forested Wetland (W)	0.97 (0.89–1.08)	**15.11**	*15.55*						
12. Tall/Medium Grassland (G)	0.67 (0.15–1.12)	**160.69**	*235.61*	1.60 (0.34–2.56)	**80.53**	*50.18*	1.09 (0.30–1.61)	**87.52**	*79.57*
13. Short Grassland (G)	0.56 (0.16–2.13)	**222.91**	*389.71*	0.72 (0.28–1.05)	**22.68**	*31.44*	0.78 (0.28–1.25)	**41.92**	*54.11*
14. Tropical Savanna (M)	0.76 (0.34–1.08)	**153.69**	*235.61*	1.47 (0.07–2.93)	**1829.5**	*1242.5*	1.14 (0.34–1.61)	**141.33**	*124.16*
15. Arid Shrubland (M)	0.48 (0.06–1.12)	**344.42**	*707.02*	0.78 (0.08–3.08)	**284.10**	*360.29*	0.47 (0.16–1.38)	**192.45**	*409.89*
16. Tropical Evergreen Forest (F)	1.85 (0.83–2.13)	**86.17**	*46.44*	2.03 (0.17–3.08)	**3459.28**	*1700.48*	1.28 (0.57–1.61)	**43.49**	*33.86*
17. Tropical Forested Wetland (W)	1.02 (1.02–1.02)	**0.27**	*0.27*	2.50 (0.98–2.93)	**136.84**	*54.6*			
18. Tropical Deciduous Forest (F)	1.32 (0.48–2.13)	**80.33**	*60.61*	1.44 (0.34–3.08)	**573.40**	*396.62*	1.43 (0.84–1.61)	**18.46**	*12.95*

19. Xeromorphic Forest/Woodland (M)	0.67 (0.19–1.12)	**53.69**	*79.70*	1.32 (0.17–3.08)	**617.83**	*466.14*	0.68 (0.21–1.61)	**95.75**	*140.34*
20. Tropical Forested Floodplain (W)				2.21 (1.67–3.08)	**33.39**	*15.121*	1.25 (1.25–1.25)	**0.32**	*0.26*
21. Desert (G)	0.26 (0.06–1.12)	**157.43**	*622.00*	0.43 (0.06–3.08)	**227.47**	*529.90*	0.44 (0.21–0.84)	**7.05**	*16.00*
22. Tropical Non-forested Wetland (W)	0.86 (0.86–0.86)	**0.24**	*0.28*	1.29 (0.68–1.50)	**6.89**	*5.36*			
23. Tropical Non-forested Floodplain (W)	1.46 (0.16–0.16)	**4.14**	*2.83*	1.67 (1.00–2.56)	**55.30**	*32.96*			
24. Temperate Non-forested Wetland (W)	0.59 (0.16–0.96)	**3.26**	*5.50*	1.90 (0.37–3.08)	**52.53**	*27.34*	0.92 (0.69–1.25)	**0.74**	*0.80*
25. Temperate Forested Floodplain (W)	0.47 (0.23–0.75)	**3.78**	*8.19*				1.06 (0.69–1.25)	**2.45**	
26. Temperate Non-forested Floodplain (W)	0.59 (0.15–0.75)	**4.62**	*7.81*				0.57 (0.28–1.25)	**1.24**	*2.16*
27. Wet Savanna (M)	0.86 (0.86–0.86)	**1.68**	*1.96*	1.20 (0.84–1.87)	**2.44**	*2.02*	1.38 (0.69–1.61)	**17.04**	*12.33*
28. Salt Marsh (W)	0.90 (0.27–1.08)	**3.80**	*4.16*	1.41 (0.84–1.99)	**1.64**	*1.16*	0.77 (0.69–0.84)	**3.13**	*4.06*

29. Mangrove (W)				1.82 (0.79–2.93)	**21.18**	*11.60*			
31. Temperate Savanna (M)	0.55 (0.07–1.90)	**341.10**	*626.84*	0.80 (0.80–0.80)	**0.23**	*0.28*	0.64 (0.38–0.80)	**39.77**	*62.22*
33. Temperate Broad-leaved Evergreen Forest (F)	1.29 (0.16–2.13)	**260.63**	*198.39*	2.05 (0.60–3.08)	**85.63**	*41.58*	0.51 (0.30–0.80)	**49.32**	*94.57*
35. Mediterranean Shrubland (M)	0.43 (0.07–0.60)	**39.62**	*93.26*				0.48 (0.33–0.81)	**25.79**	*54.13*
Zonal Σ of NH_x deposition on land	0.30 (0.02–1.06)	**2067**	*6673*	0.63 (0.02–1.77)	**3219**	*5068*	0.40 (0.07–0.87)	**474.4**	*1184*
Zonal Σ of NO_y deposition on land	0.25 (0.03–1.27)	**1817**	*6673*	0.86 (0.03–1.95)	**4421**	*5068*	0.28 (0.04–0.79)	**337.9**	*1184*
Zonal Σ of NH_x + NO_y deposition on land	0.55 (0.06–2.13)	**3885**	*6673*	1.49 (0.06–3.08)	**7640**	*5068*	0.67 (0.11–1.61)	**811.2**	*1184*

Appendix 1, Table 2. Contemporary N deposition (NH_x + NO_y, and wet plus dry deposition) onto the Defreis and Townsend land cover classification (DeFries and Townsend 1994) for Northern Hemisphere (NH) temperate latitudes, tropical latitudes and Southern Hemisphere (SH) temperate latitudes. Deposition was simulated by MOGUNTIA (Dentener and Crutzen 1994). The surface emissions of NH_3 and NO_x used in the simulations are describe in Tables 1 and 2. Nitrogen deposition onto individual biomes is summarized in Table 3. The biome classifications are grasslands – G, forests – F, cultivated lands – C, and mixed life forms – M. Zonal summarizes of NO_y, NH_x and NO_y + NH_x deposition are also included. The numbering of the vegetation types follows the numbers assigned by Defries and Townsend (1994).

Vegetation type (biome classification)	NH temperate latitudes			Tropical latitudes			SH temperate latitudes		
	Average N deposition kg N ha^{-1} (range)	N deposition Gg (10^9)	Area $ha^{-1} * 10^6$	Average N deposition kg N ha^{-1} (range)	N deposition Gg (10^9)	Area $ha^{-1} * 10^6$	Average N deposition kg N ha^{-1} (range)	N deposition Gg (10^9)	Area $ha^{-1} * 10^6$
1. Broadleaf Evergreen Forest (F)				3.34	**4475.8**	*1345*			
2. Broadleaf Deciduous Forest and Woodland (F)	6.81 (0.51–15.6)	**1093**	*156*	3.24 (0.36–9.21)	**457**	*142*	1.26 (0.45–4.16)	**41.8**	*32.9*
3. Mixed Coniferous and Broadleaf Deciduous Forest and Woodland (F)	7.59 (0.99–15.6)	**4652**	*598*	4.16 (0.39–12.5)	**4.7**	*1.1*	2.13 (0.51–4.16)	**136**	*62.2*
4. Coniferous Forest and Woodland (F)	2.34 (0.29–15.6)	**2916**	*1182*				0.89 (0.33–1.38)	**8.05**	*8.74*
5. High latitude Deciduous Forest and Woodland (F)	1.45 (0.29–6.43)	**813**	*539*						

6. & 8. Wooded C4 Grassland (M)	8.82 (3.06–13.3)	**1739**	*197*	3.98 (0.45–12.5)	**5381**	*1354*	2.70 (0.91–4.16)	**439**	*162*
7. C4 Grassland (G)	5.14 (0.66–13.3)	**986**	*190*	2.70 (0.39–12.5)	**1701**	*630*	2.14 (0.67–3.55)	**162**	*75.0*
9. Shrubs and Bare Ground (M)	2.74 (0.29–13.3)	**1047**	*380*	1.46 (0.29–8.04)	**454**	*307*	0.92 (0.24–3.55)	**381**	*414*
10. Tundra (G)	0.88 (0.29–13.3)	**269**	*293*						
11. Desert Bare Ground (G)	1.73 (0.29–8.04)	**1607**	*941*	1.45 (0.29–3.93)	**992**	*679*	1.19 (0.03–3.55)	**53.7**	*454*
12. Cultivation (C)	7.42 (0.59–15.6)	**6605**	*878*	4.15 (0.36–12.5)	**1302**	*316*	1.54 (0.47–4.16)	**212**	*136*
13. Ice (I)	0.55 (0.40–0.64)	**15.8**	*28.7*				0.05 (0.03–0.08)	**0.88**	*19.4*
14. C3 Wooded Grassland (M)	7.03 (0.59–15.6)	**1373**	*197*	3.89 (0.45–9.21)	**602**	*155*	1.93 (0.24–4.16)	**221**	*109*
15. C3 Grassland (G)	3.48 (0.34–15.6)	**3531**	*997*	2.64 (0.45–4.89)	**179**	*679*	1.15 (0.24–3.55)	**109**	*89.3*
Zonal Σ of NH_x deposition on land	1.96 (0.11–9.51)	**14150**	*6579*	1.42 (0.15–9.51)	**7063**	*4997*	0.90 (0.01–2.48)	**1070**	*1154*
Zonal Σ of NO_y deposition on land	1.80 (0.12–7.96)	1.69	**12497**	*6579* (0.11–3.75)	**8486**	*4997*	0.57 (0.02–2.09)	**694**	*1154*
Zonal Σ of NH_x + NO_y deposition on land	3.76 (0.29–15.6)	**26647**	*6579*	3.10 (0.29–12.5)	**15549**	*4997*	1.48 (0.03–4.16)	**1764**	*1154*

References

Aber JD & Driscoll CT (1997) Effects of land use, climate variation, and N deposition on N cycling and C storage in northern hardwood forests. Global Biogeochem. Cycles 11: 639–648

Aber JD, Nadelhoffer KJ, Steudler P & Melillo JM (1989) Nitrogen saturation in northern forest ecosystems. BioScience 39: 378–386

Asman WAH & Van Jaarsveld HA (1992) A variable-resolution transport model applied for NH_x in Europe. Atmospheric Environment 26A: 445–464

Asman WH (1994) Emission and deposition of ammonia and ammonium. Nova Acta Leopoldina NF 70: 263–297

Benkovitz CM, Scholtz MT, Pacyna J, Tarrason L, Dignon J, Voldner EC, Spiro PA, Jogan JA & Graedel TE (1996) Global gridded inventories of anthropogenic emission of sulfur and nitrogen. J. Geophys. Res. 101: 29,239–29,253

Bottger KA, Gravenhorst G & Ehhalt DH (1978) Atmospharische Kreislauf von Stickoxiden und Amoniak. In: Kernforschungsanlage Julich, Julich Report number 1558

Bouwman AF, Lee DS, Asman WAH, Dentener FJ, Van Der Hoek KW & Olivier JGJ (1997) A global high-resolution emission inventory for ammonia. Global Biogeochem. Cycles 11: 561–587

Carpenter S, Caraco NF, Corrlee DL, Howarth RW, Sharpley AN & Smity VH (1998) Nonpoint pollution of surface waters with phosphorus and nitrogen. Ecol. App. 8: 559–568

Chameides WL, Kasibhatla PS, Yienger J & Levy II H (1994) Growth of continental-scale metro-agro-plexes, regional ozone pollution, and world food production. Science 264: 74–77

Cleveland CC, Townsend AR, Schimel DS, Howarth RW, Hedin LO, Perakis SS, von Fischer JD, Wasson MF, Latty EF & Elseroad A (submitted) Global patterns of terrestrial biological concentration (N_2) fixation in natural ecosystems. Global Biogeochem. Cycles

Cornell S, Rendell A & Jickells T (1995) Atmospheric inputs of dissolved organic nitrogen to the ocean. Nature 376: 243–246

Cramer W, Claussen M & Solomon AM (1995) An assessment of different climate change scenarios for the global redistribution of agricultural land. First Science Conference, Int. Geosphere-Biosphere Programme, Global Anal., Interpret., and Model, Garmisch-Partenkirchen, Germany

Crutzen PJ & Andreae MO (1990) Biomass burning in the tropics: Impact on atmospheric chemistry and biogeochemical cycles. Science 250. 1669–1678

Crutzen PJ & Gidel LT (1983) A two-dimensional photochemical model of the atmosphere 2: The tropospheric budges of the anthropogenic chlorocarbons CO, CH_4, CH_3Cl and the effect of the various NO_x sources on tropospheric ozone. J. Geophys. Res. 88: 6641–6661

Crutzen PJ & Zimmermann PH (1991) The changing photochemistry of the troposphere. Tellus 43AB: 136–151

Davidson EA & Kingerlee W (1997) Global inventory of nitric oxide emissions from soils. Nutrient Cycling in Agroecosystems 48: 37–50

Dawson GA (1977) Atmospheric amonia from undisturbed land. J. Geophys. Res. 82: 3125–3133

DeFries RM & Townshend JRG (1995) An initial coarse resolution NDVI-derived global land cover classification. In ISLSCP Initiative 1: Global Data Sets for Land/Atmosphere Models 1987–1988. CD-ROM NASA, Greenbelt, Md

Delmas R, Derca D & Jambert C (1997) Global inventory of NO_x sources. Nutrient Cycling in Agroecosystems 48: 51–60

Newell RE, Kidson JW, Vincent DG & Boer GJ (1974) The General Circulation of the Tropical Atmosphere and Interactions with Extra Tropical Latitudes. MIT Press, Cambridge, Mass

Olivier JGJ, Bouwman AF, van der Maas CWM, Berdowski JJM, Veldt C, Bloos JPJ, Visschedijk AJH, Zandveld PYJ & Havelag JL (1995) Description of EDGAR Version 2.0: A set of global emission inventories of greenhouse gases and ozone-depleting substances for all anthropogenic and most natural sources on a per country basis on 1 × grid. In: RIWM, 771060 002

Oort AH (1983) Global atmospheric circulation statistics, 1958–1973. In: Natl. Oceanic and Atmos. Admin., Report Number 14

Penner JE, Atherton CS, Dignon J, Ghan SJ & Walton JJ (1991) Tropospheric nitrogen: a three dimensional study of sources, distributions and deposition. J. Geophys. Res. 96D: 959–990

Penner JE, Atherton CS & Graedel TE (1994) Global emissions and models of photochemically active compounds. In: Prinn RG (Ed.) Global Atmospheric-Biospheric Chemistry, Plenum Press, New York

Potter CS, Matson PA, Vitousek PM & Davidson EA (1996) Process modeling of controls on nitrogen trace gas emissions from soils worldwide. J. Geophys. Res. 101: 1361–1377

Prather M, Derwent R, Ehhalt D, Fraser P, Sanhueza E & Zhou X (1995) Other Trace Gases and Atmospheric Chemistry. In: Climate Change 1994, Radiative Forcing of Climate Change. Intergovernmental Panel on Climate Change (pp 73–126). Cambridge University Press, Cambridge

Price C, Penner J & Prather M (1997a) NO_X from lightning, 1, Global distribution based on lightning physics. J. Geophys. Res. 102: 5929–5941

Price C, Penner J & Prather M (1997b) NO_X from lightning, 2, Constraints from the global atmospheric electric circuit. J. Geophys. Res. 102: 5943–5951

Prospero JM, Barrett K, Church T, Dentener F, Duce RA, Galloway JN, Levy II H, Moody J & Quinn P (1996) Atmospheric deposition of nutrients to the North Atlantic Basin. Biogeochemistry 35: 27–73

Rendell AR, Ottley CJ, Jickells TD & Harrison RM (1993) The atmospheric input of nitrogen species to the North Sea. Tellus 45B: 53–63

Ridley BA & Atlas E (in press, anticipated publication January 1999) Nitrogen Compounds. In: Brasseur GP, Orlando JJ & Tyndall GS (Eds) Atmospheric Chemistry and Global Change. Oxford University Press, New York

Ridley BA, Dye JE, Walega JG, Zhenge J, Grahek FE & Rison W (1996) On the production of active nitrogen by thunderstorms over New Mexico. J. Geophys. Res. 101: 20,985–21,005

Sanhueza E, Fraser PJ & Zander RJ (1995) Scientific assessment of ozone depletion 1994: Source gasses: Trends and budgets. In: World Meterological Organization, 37

Schimel D, Braswell BH, McKeown R, Ojima DS, Parton WJ & Pulliam W (1996) Climate and nitrogen controls on the geography and timescales of terrestrial biogeochemical cycling. Global Biogeochem. Cycles 10: 677–692

Schimel DS, Braswell BH & Parton WJ (1997) Equilibration of the terrestrial water, nitrogen, and carbon cycles. Proceedings of the National Academy of Sciences 94: 8280–8283

Schlesinger WH & Hartley AE (1992) A global budget for atmospheric NH_3. Biogeochemistry 15: 191–211

Schulze ED (1989) Air pollution and forest decline in a spruce (Picea Abies) forest. Science 244: 776–783

Setzer AW & Pereira MC (1991) Amazonia biomass burnings in 1987 and an estimate of their tropospheric emissions. Ambio 20: 19–22

Soderlund R & Svensson BH (1976) Their global nitrogen cycle. Nitrogen, Phosphorous and Sulphur Global Cycles SCOPE 7: 23–73

Stedman DH & Shetter RE (1983) The global budget of atmospheric nitrogen species. In: Schwartz SE (Ed.) Trace Atmospheric Constituents (pp 411–454). John Wiley & Sons, New York

Stohl A, Williams E, Wotawa G & Kromp-Kolb H (1996) A European inventory of soil nitric oxide emissions and the effect of these emissions on the photochemical formation of ozone. Atmos. Env. 30: 3,741–3,755

Sulzman J, Braswell BH, Holland EA & Lamarque JF (1997) Poster: A Comprehensive Compilation of Atmospheric Deposition Data for Global and Regional Studies. Ecological Society of America, Albuquerque, NM

Townsend AR, Braswell BH, Holland EA & Penner JE (1996) Spatial and temporal patterns in potential terrestrial carbon storage resulting from deposition of fossil fuel derived nitrogen. Ecological Applications 6: 806–814

Van Breemen N, Bottough PA, Velthorst EJ, Van Dobben HF, De Wit T, Ridder TB & Reunders HF (1982) Soil acidification from atmospheric ammonium sulfate in forest canopy throughfall. Nature 299: 548–550

Veldkamp E & Keller M (1997) Fertilizer induced nitric oxide emissions from agricultural soils. Nutrient Cycling in Agroecosystems 48: 69–77

Vitousek PM, Aber JD, Howarth RW, Likens GE, Matson PA, Schindler DW, Schlesinger WH & Tilman DG (1997) Human alteration of the global nitrogen cycle: sources and consequences. Ecol. Appl. 7: 737–750

Wagenbach D, Munnich KO, Schotterrer U & Aeschger H (1988) The anthropogenic impact on snow chemistry at Colle Gnifetti, Swiss Alps. Ann. Glaciol. 10: 183–187

Warneck P (1988) Chemistry of the Natural Atmosphere. Academic Press, San Diego

Wesely ML (1989) Parameterization of surface resistance to gaseous dry deposition in regional-scale numerical models. Atmospheric Environment 23: 1293–1304

Whelpdale DM, Dorling SR, Hicks BB & Summers PW (1996) Atmospheric Processes. In: World Meterological Organizational Global Atmosphere Watch: Global Acid Deposition Assessment, 106

Whelpdale DM & Galloway JN (1994) Sulphur and nitrogen oxide fluxes in the North Atlantic atmosphere. Global Biogeochem. Cycles 8: 481–493

Whelpdale DM, Summers PW & Sanhueza E (1997) A global overview of atmospheric acid deposition fluxes. Environmental Monitoring and Assessment 48: 217–227

Williams EJ, Guenther A & Fehsenfeld FC (1992) An inventory of nitric oxide emissions from soils in the United States. J. Geophys. Res. 97: 7511–7519

Yienger JJ & Levy II H (1995) Empirical model of global soil-biogenic NO_X emissions. J. Geophys. Res. 100: 11,447–11,464

Zimmermann PH (1987) MOGUNTIA: A handy global tracer model. Proceedings of the Sixteenth NATO/CCMS International Technical Meeting on Air Pollution Modeling and Its Application, New York

Biogeochemistry **46:** 45–65, 1999.

Nitrogen stable isotopic composition of leaves and soil: Tropical versus temperate forests

L.A. MARTINELLI[1], M.C. PICCOLO[1], A.R. TOWNSEND[2], P.M. VITOUSEK[3], E. CUEVAS[4], W. McDOWELL[5], G.P. ROBERTSON[6], O.C. SANTOS[7] & K. TRESEDER[3]

[1]*Cena – Av. Centenário 303, Piracicaba-SP. 13416-000, Brazil;* [2]*INSTAAR and Department of EPO-Biology, Campus Box 450, University of Colorado, Boulder, CO 80309, U.S.A.;* [3]*Department of Biological Sciences, Stanford University, Stanford, CA 94305-5020, U.S.A.;* [4]*IVIC, BAMCO CCS-199-00, PO Box 025322, Miami, Fl 33102-5322, U.S.A.;* [5]*Department of Natural Resources, University of New Hampshire, Durham, NH, 03824, U.S.A.;* [6]*W.K. Kellogg Biological Station, Michigan State University, Hickory Corners, MI 9060-9516, U.S.A.;* [7]*Meteorological Institute, Ministry of Science Technology and Environment, Aptdo 17032, Habana 17, CP 11700, Havana, Cuba*

Received 10 December 1998

Key words: N15, nitrogen, nutrient cycling, plants, stable isotopes, soil, temperate forest, tropical forest

Abstract. Several lines of evidence suggest that nitrogen in most tropical forests is relatively more available than N in most temperate forests, and even that it may function as an excess nutrient in many tropical forests. If this is correct, tropical forests should have more open N cycles than temperate forests, with both inputs and outputs of N large relative to N cycling within systems. Consequent differences in both the magnitude and the pathways of N loss imply that tropical forests should in general be more ^{15}N enriched than are most temperate forests. In order to test this hypothesis, we compared the nitrogen stable isotopic composition of tree leaves and soils from a variety of tropical and temperate forests. Foliar $\delta^{15}N$ values from tropical forests averaged 6.5‰ higher than from temperate forests. Within the tropics, ecosystems with relatively low N availability (montane forests, forests on sandy soils) were significantly more depleted in ^{15}N than other tropical forests. The average $\delta^{15}N$ values for tropical forest soils, either for surface or for depth samples, were almost 8‰ higher than temperate forest soils. These results provide another line of evidence that N is relatively abundant in many tropical forest ecosystems.

Introduction

A number of lines of evidence suggest that N in most tropical forests is relatively more available than is N in most temperate forests. On average, more N circulates annually through lowland tropical forests, and does so at higher concentrations, than through temperate forests (Proctor et al. 1983;

Vitousek 1984; Vitousek & Sanford 1986; Vogt et al. 1986). Emissions of N-containing trace gases are also higher, both absolutely and as a fraction of N circulating through forests (Keller et al. 1986, 1993; Matson & Vitousek 1987, 1990). Comparable data on rates of N mineralization and leaching losses are sparser, but they generally show greater rates of N cycling and loss in many lowland tropical forests (Vitousek & Denslow 1986; Lewis 1986; Matson et al. 1987; Neill et al. 1995). Overall, these observations suggest that N functions as an excess nutrient in most tropical forests, but not in the majority of temperate forests. The major exceptions to this generalization in the tropics are forests on white-sand soils and montane tropical forests; by the measures above, N is in relatively short supply in these ecosystems (Salati et al. 1982; Cuevas & Medina 1988; Tanner et al., in press). In the temperate zone, the major exceptions are forests dominated by symbiotic N fixing trees (usually monocultures), and forests that receive substantial anthropogenic N deposition (Binkley et al. 1992; Aber et al. 1995; Berendse et al. 1993).

If N functions as an excess nutrient in tropical forests, the N cycle in such systems should be more open than in temperate forests, with both inputs and outputs of N large relative to internal N cycling. Moreover, the pathways of N loss should differ: losses from low-N systems may be predominantly in the form of DON (Hedin et al. 1995), while leaching losses of nitrate and nitrification/denitrification driven trace gas fluxes should predominate where N is in excess (Matson et al., this volume). These differences in both the magnitude and the pathways of loss imply that tropical forests should in general contain N that is more enriched in ^{15}N than most temperate forests. This relative enrichment should occur because pathways of loss in N-rich systems are more likely to be fractionating, and because losses by fractionating pathways leave the N remaining within the system enriched (Hogberg 1997). In order to test this hypothesis, we compared the nitrogen stable isotopic composition of tree leaves and soils from a variety of tropical and temperate forests.

Methods

We surveyed N stable isotopic composition and N content of leaves from adult trees of non-leguminous species from temperate and tropical forests. Where available, the same data for soil organic matter also were evaluated. As there is significant variation with depth in both concentration and stable isotopic composition of nitrogen in soils, data were grouped according to depth. Surface samples were those collected from no more than 10 cm deep, and samples collected below this depth were called depth samples. The nitrogen stable isotopic composition was expressed as δ (‰) notation:

$$\delta^{15}\text{N} = (\text{R}_{\text{sample}}/\text{R}_{\text{std}} - 1) \cdot 1000,$$

where R is the ratio of $^{15}N/^{14}N$ of the sample and standard (std). The isotopic standard for nitrogen is atmospheric air. We predicted that tropical forests would have higher ^{15}N values in relation to temperate forests; statistical differences were tested with a *t*-test for unequal variance.

Results and discussion

Tropical vs temperate systems

The average $\delta^{15}N$ value for tropical foliage was 3.7±3.5‰ ($n = 73$), which is significantly greater ($p < 0.01$) than the temperate forest value of −2.8±2.0‰ ($n = 90$) [Table 1, Figure 1]. This 6.5‰ difference occurred despite the fact that trees from low-N montane and white-sand tropical sites were included in the analysis. If these are excluded, the average $\delta^{15}N$ value for tropical trees increases to 4.7±2.1‰ ($n = 65$). The average concentration of nitrogen in leaves from tropical forests was 1.9±0.8‰ ($n = 78$), which was not significantly different from the 1.6±0.5‰ ($n = 28$) average value found for temperate forests. When data are grouped by site, there is a significant positive correlation between ^{15}N and N concentration ($p < 0.007$); sites with higher nitrogen concentration in their leaves tend to have higher ^{15}N values (Figure 2).

Tropical forest soils were also much more ^{15}N-enriched than temperate forest soils (Table 2, Figure 3a), with tropical ^{15}N values averaging almost 8‰ higher, both at the surface and at depth. Variation with depth followed the classical pattern discussed by Nadelhoffer and Fry (1988); the average ^{15}N value for depth samples was approximately 2‰ higher than for surface samples in both tropical and temperate forests (Figure 3a).

The comparison of nitrogen concentration of soil organic matter between tropical and temperate forests was difficult to make due to the small number of samples from temperate forest soils. With the available data, the average N concentration was smaller in the tropical soils than in the temperate forests for both surface and depth samples (Figure 3b).

The results clearly showed that the ^{15}N values of tree leaves and soil organic matter were significantly higher in tropical forests than in temperate forests. Therefore, the initial working hypothesis was confirmed, in that the results are consistent with tropical forests having a more open nitrogen cycle, with greater losses via fractionating pathways, suggesting that N is in relative excess in many moist tropical forests.

Table 1. $\delta^{15}N$ (‰) values of plant species. %N is the nitrogen concentration (%).

Species	Site	Region	Country	^{15}N	% N	Ref
Metrosideros polymorpha	Thurston	Hawaii	U.S.A.	−6.8	0.87	1
Metrosideros polymorpha	Olaa	Hawaii	U.S.A.	−4.9	1.12	1
Metrosideros polymorpha	Laupahoehoe	Hawaii	U.S.A.	+0.9	1.42	1
Metrosideros polymorpha	Kohala	Hawaii	U.S.A.	−2.2	1.14	1
Metrosideros polymorpha	Malokai	Hawaii	U.S.A.	−2.3	1.06	1
Metrosideros polymorpha	Kauai	Hawaii	U.S.A.	−0.5	0.86	1
Cibotium glaucum	Thurston	Hawaii	U.S.A.	−9.3	1.79	1
Cibotium glaucum	Olaa	Hawaii	U.S.A.	−6.0	1.53	1
Cibotium glaucum	Laupahoehoe	Hawaii	U.S.A.	+0.7	1.82	1
Cibotium glaucum	Kohala	Hawaii	U.S.A.	−3.7	1.79	1
Cibotium glaucum	Malokai	Hawaii	U.S.A.	−1.3	1.79	1
Cibotium glaucum	Kauai	Hawaii	U.S.A.	−1.3	1.47	1
Amphirrox latifolia	Samuel	Rondônia	Brazil	+6.8	3.16	2
Licania hispidula	Samuel	Rondônia	Brazil	+5.7	1.15	2
Maquira guianensis	Samuel	Rondônia	Brazil	+6.4	1.06	2
Naucleopsis sp	Samuel	Rondônia	Brazil	+8.3	3.51	2
Naucleopsis sp	Samuel	Rondônia	Brazil	+6.5	0.85	2
Neea sp	Samuel	Rondônia	Brazil	+9.6	3.76	2
Protium carnosum	Samuel	Rondônia	Brazil	+5.7	0.98	2
Protium sp	Samuel	Rondônia	Brazil	+4.4	1.12	2
Protium sp	Samuel	Rondônia	Brazil	+5.7	1.10	2
Tachigalia cavipes	Samuel	Rondônia	Brazil	+7.4	3.02	2
Undetermined	Samuel	Rondônia	Brazil	+6.4	3.96	2
Undetermined	Samuel	Rondônia	Brazil	+6.3	1.04	2
Mixture	Faz. Nova Vida	Rondônia	Brazil	+8.0	1.43	2
Rinorea racemosa	Varzea	rio Amazonas	Brazil	+2.0	1.00	2
Inga sp	Varzea	rio Amazonas	Brazil	+1.2	2.22	2
Oxandra polyantha	Varzea	rio Amazonas	Brazil	+2.5	2.74	2
Nectandra amazonum	Varzea	rio Amazonas	Brazil	+3.7	2.76	2
Leonia racemosa	Varzea	rio Amazonas	Brazil	+4.7	2.80	2
Undetermined	Varzea	rio Amazonas	Brazil	+4.0	1.22	2
Sacoglotis sp	Varzea	rio Amazonas	Brazil	+1.8	1.22	2
Ficus glabrata	Varzea	rio Amazonas	Brazil	+1.3	1.77	2
Tichila sp	Varzea	rio Amazonas	Brazil	+1.2	0.29	2
Laetia crynbulosa	Varzea	rio Amazonas	Brazil	+2.4	2.76	2
Undetermined	Campina	Manaus	Brazil	−2.2	1.86	3
Undetermined	Campina	Manaus	Brazil	−0.2	2.49	3
Undetermined	Campina	Manaus	Brazil	+0.6	1.30	3
G.thophilurum	Campina	Manaus	Brazil	−7.0		4
Sapotaceae	Campina	Manaus	Brazil	−7.0		4
Lecitidaceae	R.Ducke	Manaus	Brazil	+5.8		4
Virola surinasuensis	R. Ducke	Manaus	Brazil	+5.9		4
Carapa guianensis	R. Ducke	Manaus	Brazil	+7.0		4
Theobroma cacau	R. Ducke	Manaus	Brazil	+6.2	1.32	5

Table 1. Continued.

Species	Site	Region	Country	^{15}N	% N	Ref
Melastoma bellucia	R. Ducke	Manaus	Brazil	+3.3	2.38	5
Melastoma bellucia	R. Ducke	Manaus	Brazil	+4.6	1.68	5
Melastoma bellucia	R. Ducke	Manaus	Brazil	+4.7	1.81	5
Cecropia eucomona	R. Ducke	Manaus	Brazil	+5.9	2.20	5
Cecropia eucomona	R. Ducke	Manaus	Brazil	+4.9	1.86	5
Piper piperaceae	R. Ducke	Manaus	Brazil	+4.6	2.58	5
Aegiphula scandens	R. Ducke	Manaus	Brazil	+5.6	3.09	5
Rolinea exsucca	R. Ducke	Manaus	Brazil	+2.0	1.48	5
Verbenaceae	R. Ducke	Manaus	Brazil	+6.0	2.12	5
Alchornea schomburgkii	R. Ducke	Manaus	Brazil	+1.7	2.10	5
Myrtaceae	Cerrado	Brasìlia	Brazil	+1.8	0.94	6
Compositae	Cerrado	Brasìlia	Brazil	+0.1	1.31	6
Rubiaceae	Cerrado	Brasìlia	Brazil	−0.6	1.09	6
Melastomataceae	Cerrado	Brasìlia	Brazil	+1.3	1.30	6
Flacourtia rukan	Toong−fax		Thailand	+4.5	1.70	5
Anacardium occidentale	Toong−fax		Thailand	+3.6	1.30	5
Zizyplus mauritana	Toong−fax		Thailand	+2.5	1.83	5
Terminalia catappa	Kwae		Thailand	+7.5	1.88	5
Lansium domesticum	Nakhon Pathom		Thailand	+6.0	2.18	5
Ficus glberrima	Kwae		Thailand	+9.4	1.78	5
Crataeva erythrocarpa	Kwae		Thailand	+8.0	2.44	5
Kerangas scrub			Sarawak	−2.3		7
Kerangas scrub			Sarawak	−7.4		7
Kerangas forest			Sarawak	−2.4		7
Kerangas forest			Sarawak	−3.5		7
8 ECM species	Korup	Cameroon	Africa	+4.9	2.10	20
10 VAM species	Korup	Cameroon	Africa	+4.6	2.11	20
Quercus kelloggii	Mix Canyon	California	U.S.A.	+0.4		8
Pinus sabiniana	Mix Canyon	California	U.S.A.	+0.8		8
Pinus contorta ssp.	Grass Lake	California	U.S.A.	+0.2		8
Pinus albicaulis	Carson Pass-3	California	U.S.A.	−0.3		8
Pinus contorta ssp.	Carson Pass-4	California	U.S.A.	+0.4		8
Tsuga mertensiana	Carson Pass-5	California	U.S.A.	+1.0		8
Abies concolor	Rice canyon	California	U.S.A.	−0.8		8
Pinus ponderosa	Rice canyon	California	U.S.A.	−0.5		8
Prunus emarginata	Rice canyon	California	U.S.A.	−0.5		8
Salix scouleriana	Rice canyon	California	U.S.A.	−0.6		8
Picea abies	Fichtelgebirge	Bavaria	Germany	−3.1	1.75	9
Picea abies	Fichtelgebirge	Bavaria	Germany	−3.5	1.70	9
Picea abies	Fichtelgebirge	Bavaria	Germany	−3.6	1.70	9
Picea abies	Fichtelgebirge	Bavaria	Germany	−3.7	1.50	9
Picea abies	Fichtelgebirge	Bavaria	Germany	−3.3	1.50	9
Picea abies	Fichtelgebirge	Bavaria	Germany	−3.8	1.45	9

Table 1. Continued.

Species	Site	Region	Country	^{15}N	% N	Ref
Picea abies	Fichtelgebirge	Bavaria	Germany	−3.8	1.50	9
Picea abies	Fichtelgebirge	Bavaria	Germany	−4.0	1.45	9
Picea abies	Fichtelgebirge	Bavaria	Germany	−3.9	1.40	9
Picea abies	Fichtelgebirge	Bavaria	Germany	−3.9	1.40	9
Picea abies	Fichtelgebirge	Bavaria	Germany	−3.9	1.20	9
Picea abies	Fichtelgebirge	Bavaria	Germany	−3.4	1.48	9
Picea abies	Fichtelgebirge	Bavaria	Germany	−3.8	1.52	9
Picea abies	Fichtelgebirge	Bavaria	Germany	−3.9	1.54	9
Picea abies	Fichtelgebirge	Bavaria	Germany	−4.0	1.48	9
Salix purpurea	Col d'Ornon	Northern Alps	France	−3.8		10
Betula verrucosa	Col d'Ornon	Northern Alps	France	−6.0		10
Salix purpurea	Col d'Ornon	Northern Alps	France	−3.4		10
Acer pseucoplatanus	Col d'Ornon	Northern Alps	France	−4.3		10
Acer pseudoplatanus	Col d'Ornon	Northern Alps	France	−3.6		10
Corylus avellana	Col d'Ornon	Northern Alps	France	−4.4		10
Fraxinus excelsior	Col d'Ornon	Northern Alps	France	−3.8		10
Abies alba	Col d'Ornon	Northern Alps	France	−3.6		10
Rubus sp	Col d'Ornon	Northern Alps	France	−4.3		10
Spruce-fir	Great Smoky Mt.	Tennessee	U.S.A.	−2.2		11
Beech	Great Smoky Mt.	Tennessee	U.S.A.	−0.8		11
Cove hardwood	Great Smoky Mt.	Tennessee	U.S.A.	−1.9		11
Mixed hardwood	Great Smoky Mt.	Tennessee	U.S.A.	−1.0		11
Floodplain poplar	Great Smoky Mt.	Tennessee	U.S.A.	−2.4		11
Yellow poplar	Great Smoky Mt.	Tennessee	U.S.A.	−3.5		11
Pine	Great Smoky Mt.	Tennessee	U.S.A.	−1.7		11
Xeric oak	Great Smoky Mt.	Tennessee	U.S.A.	−1.8		11
Picea glauca	Dalton Highway	Alaska	U.S.A.	−6.2	0.97	12
Picea mariana	Dalton Highway	Alaska	U.S.A.	−10.1	0.91	12
Acer rubrum	Walker Branch	Tennessee	U.S.A.	−3.5		13
Cornus florida	Walker Branch	Tennessee	U.S.A.	−3.9		13
Liliodendron tulipifera	Walker Branch	Tennessee	U.S.A.	−3.4		13
Acer rubrum	Walker Branch	Tennessee	U.S.A.	−1.7		13
Cornus florida	Walker Branch	Tennessee	U.S.A.	−1.0		13
Liliodendron tulipifera	Walker Branch	Tennessee	U.S.A.	−2.4		13
Facus grandifolia	Bear Brook	Maine	U.S.A.	−0.7	2.30	14
Acer spp.	Bear Brook	Maine	U.S.A.	−1.7	2.30	14
Betula alleghaniensis	Bear Brook	Maine	U.S.A.	−1.1	2.20	14
Picea rubens	Bear Brook	Maine	U.S.A.	−0.6	1.10	14
Picea abis	Southern Sweden	Sweden	Sweden	−2.2	1.30	15
Juniperus communis	Subsite 1&2	Scotland	Scotland	0.6	1.70	16
Juniperus communis	Subsite 3 (boggy)	Scotland	Scotland	−5.3	1.00	16
Betula nana		Northern Sweden	Sweden	−7.4	1.90	17
Betula nana		Northern Sweden	Sweden	−3.9	3.30	17
Pinus sylvestris		Northern Sweden	Sweden	−3.9		17

Table 1. Continued.

Species	Site	Region	Country	^{15}N	% N	Ref
Picea abies		Central Sweden	Sweden	−1.3		18
Pseudotsuga menziesii	Andrews	Oregon	U.S.A.	−3.2		19
Acer rubrum	Harvard Forest	Massachusetts	U.S.A.	−3.6		19
Quercus rubra	Harvard Forest	Massachusetts	U.S.A.	−2.7		19
Tsuga canadensis	Harvard Forest	New Hampshire	U.S.A.	−3.7		19
Acer rubrum	Harvard Forest	New Hampshire	U.S.A.	−5.9		19
Pinus strobus	Harvard Forest	New Hampshire	U.S.A.	−1.4		19
Betula papyrifera	Harvard Forest	New Hampshire	U.S.A.	−0.6		19
Facus grandifolia	Harvard Forest	New Hampshire	U.S.A.	−1.2		19
Prunus pensylvanica	Harvard Forest	New Hampshire	U.S.A.	−0.4		19
Betula lutea	Harvard Forest	New Hampshire	U.S.A.	−2.7		19
Pinus spp	North Inlet	South Carolina	U.S.A.	−0.9		19
Pinus resinosa	North Lakes	Wiscosin	U.S.A.	−2.0		19
Quercus rubra	North Lakes	Wiscosin	U.S.A.	−3.0		19
Betula papyrifera	North Lakes	Wiscosin	U.S.A.	−2.5		19
Picea glauca	Bonanza creek	Alaska	U.S.A.	−3.3		19
Populus tremuliodes	Bonanza creek	Alaska	U.S.A.	−1.4		19
Fraxinus spp	Cedar Creek	Minnesota	U.S.A.	−4.2		19
Betula papyrifera	Cedar Creek	Minnesota	U.S.A.	−4.0		19
Acer rubrum	Cedar Creek	Minnesota	U.S.A.	−5.2		19
Quercus macrocarpa	Cedar Creek	Minnesota	U.S.A.	−4.5		19
Quercus rubra	Cedar Creek	Minnesota	U.S.A.	−3.3		19
Cornus florida	Coweeta	North Carolina	U.S.A.	−1.8		19
Acer rubrum	Coweeta	North Carolina	U.S.A.	−5.2		19
Liliodendron tulipifera	Coweeta	North Carolina	U.S.A.	−5.3		19

1 – Vitousek PM (nonpublished data); 2 – Almeida S (1995); 3 – McClain M (nonpublished data); 4 – Salati et al. (1982) 5 – Yoneyama et al (1993); 6 – Sprent et al. (1996); 7 – Treseder K (nonpublished data); 8 – Virginia and Delwiche (1982); 9 – Gebauer and Schulze (1991); 10 – Domenach et al. (1989); 11 – Garten Jr and Miegroet (1994); 12 – Schulze et al. (1994); 13 – Garten Jr (1993); 14 – Adelhoffer et al. (1995); 15 – Nasholm et al. (1997); 16 – Hill et al. (1996); 17 – Michelsen et al. (1996); 18 – Hogberg et al. (1996); 19 – Fry (1991); 20 – Hogberg and Alexander (1995)

Nutrient rich vs nutrient poor systems in the tropics

Forests on white-sand soils in the tropics are considered to be nitrogen-poor systems (Vitousek & Sanford 1986; Cuevas & Medina 1988). Table 3 summarizes the elemental composition of leaves in contrasting vegetation types in the Amazon. Higher N concentrations were found in Varzea forest, followed by Terra-firme forests in Samuel and Manaus. The same trend was also found by Furch and Klinge (1989). As expected, the lowest concentrations were in the Campina forests. Following our initial hypothesis,

Table 2. ^{15}N values (‰) for soil samples. Z is the soil depth (cm), %N is the nitrogen concentration (%) and Re stands for references.

Country	Region	Soil type	*z*	^{15}N	%N	Re
Brazil	Pará	Yellow latosol (Hapludox)	0–5	+9.8	0.26	1
Brazil	Pará	Yellow latosol (Hapludox)	5–10	+10.8	0.17	1
Brazil	Pará	Yellow latosol (Hapludox)	10–20	+12.0	0.11	1
Brazil	Pará	Yellow latosol (Hapludox)	20–30	+12.5	0.09	1
Brazil	Pará	Yellow latosol (Hapludox)	30–40	+12.6	0.07	1
Brazil	Pará	Yellow latosol (Hapludox)	40–50	+13.2	0.05	1
Brazil	Pará	Yellow latosol (Hapludox)	50–60	+12.8	0.04	1
Brazil	Pará	Yellow latosol (Hapludox)	60–70	+12.9	0.03	1
Brazil	Pará	Yellow latosol (Hapludox)	100–110	+12.4	0.02	1
Brazil	Pará	Yellow latosol (Hapludox)	140–150	+11.9	0.01	1
Brazil	Pará	Yellow latosol (Hapludox)	0–5	+8.4	0.13	1
Brazil	Pará	Yellow latosol (Hapludox)	5–10	+9.2	0.05	1
Brazil	Pará	Yellow latosol (Hapludox)	10–20	+9.4	0.04	1
Brazil	Pará	Yellow latosol (Hapludox)	20–30	+9.9	0.02	1
Brazil	Pará	Yellow latosol (Hapludox)	30–40	+9.8	0.01	1
Brazil	Pará	Yellow latosol (Hapludox)	40–50	+9.3	0.01	1
Brazil	Pará	Yellow latosol (Hapludox)	50–60	+9.3	0.00	1
Brazil	Pará	Yellow latosol (Hapludox)	60–70	+8.5	0.01	1
Brazil	Pará	Yellow latosol (Hapludox)	80–90	+9.8	0.00	1
Brazil	Pará	Yellow latosol (Hapludox)	120–130	+8.6	0.00	1
Brazil	Amazonas	Yellow latosol (Hapludox)	0–3	+7.7	0.38	1
Brazil	Amazonas	Yellow latosol (Hapludox)	3–12	+9.8	0.19	1
Brazil	Amazonas	Yellow latosol (Hapludox)	12–36	+11.9	0.09	1
Brazil	Amazonas	Yellow latosol (Hapludox)	36–51	+14.9	0.02	1
Brazil	Amazonas	Yellow latosol (Hapludox)	51–80	+15.6	0.01	1
Brazil	Amazonas	Yellow latosol (Hapludox)	80–140	+20.0	0.00	1
Brazil	Amazonas	Yellow latosol (Hapludox)	140	+21.7	0.00	1
Brazil	Rondônia	Red-yellow latosol (Hapludox)	0–5	+9.8	0.24	1
Brazil	Rondônia	Red-yellow latosol (Hapludox)	5–10	+10.1	0.22	1
Brazil	Rondônia	Red-yellow latosol (Hapludox)	10–20	+10.5	0.16	1
Brazil	Rondônia	Red-yellow latosol (Hapludox)	20–30	+10.8	0.13	1
Brazil	Rondônia	Red-yellow podzolic latosol (Kandiudult)	0–5	+8.6	0.18	1
Brazil	Rondônia	Red-yellow podzolic latosol (Kandiudult)	5–10	+9.2	0.14	1
Brazil	Rondônia	Red-yellow podzolic latosol (Kandiudult)	10–20	+9.9	0.08	1
Brazil	Rondônia	Red-yellow podzolic latosol (Kandiudult)	20–30	+10.3	0.08	1
Brazil	Rondônia	Red-yellow podzolic (Paleudult)	0–5	+10.7	0.21	1
Brazil	Rondônia	Red-yellow podzolic (Paleudult)	5–10	+11.4	0.14	1
Brazil	Rondônia	Red-yellow podzolic (Paleudult)	10–20	+11.9	0.11	1
Brazil	Rondônia	Red-yellow podzolic (Paleudult)	20–30	+11.9	0.09	1
Brazil	Rondônia	Yellow latosol (Hapludox)	0–5	+9.3	0.28	1
Brazil	Rondônia	Yellow latosol (Hapludox)	5–10	+9.8	0.19	1
Brazil	Rondônia	Yellow latosol (Hapludox)	10–20	+11.3	0.12	1

Table 2. Continued.

Country	Region	Soil type	*z*	^{15}N	%N	Re
Brazil	Rondônia	Yellow latosol (Hapludox)	20–30	+11.7	0.08	1
Brazil	Rondônia	Yellow latosol (Hapludox)	0–5	+6.4	0.09	1
Brazil	Rondônia	Yellow latosol (Hapludox)	5–10	+7.6	0.07	1
Brazil	Rondônia	Yellow latosol (Hapludox)	10–20	+9.2	0.05	1
Brazil	Rondônia	Yellow latosol (Hapludox)	20–30	+10.6	0.04	1
Brazil	Rondônia	Red yellow podzolic latosol (Kandiudult)	0–5	+11.2	0.12	1
Brazil	Rondônia	Red yellow podzolic latosol (Kandiudult)	5–10	+12.2	0.10	1
Brazil	Rondônia	Red yellow podzolic latosol (Kandiudult)	10–20	+13.3	0.06	1
Brazil	Rondônia	Red yellow podzolic latosol (Kandiudult)	20–30	+13.6	0.04	1
Brazil	Rondônia	Red yellow podzolic latosol (Kandiudult)	30–40	+13.1	0.04	1
Brazil	Rondônia	Red yellow podzolic latosol (Kandiudult)	40–50	+12.9	0.04	1
Brazil	Rondônia	Red yellow podzolic latosol (Kandiudult)	80–90	+12.4	0.06	1
Brazil	Rondônia	Red yellow podzolic latosol (Kandiudult)	120–130	+12.7	0.04	1
Brazil	Rondônia	Red-yellow podzolic (Paleudult)	0–10	+10.5	0.07	1
Brazil	Rondônia	Red-yellow podzolic (Paleudult)	10–20	+11.7	0.04	1
Brazil	Rondônia	Red-yellow podzolic (Paleudult)	20–30	+11.9	0.02	1
Brazil	Rondônia	Red-yellow podzolic (Paleudult)	30–40	+12.1	0.02	1
Brazil	Rondônia	Red-yellow podzolic (Paleudult)	40–50	+12.2	0.01	1
Brazil	Rondônia	Red-yellow podzolic (Paleudult)	70–80	+11.4		1
Brazil	Rondônia	Red-yellow podzolic (Paleudult)	90–100	+9.8		1
Brazil	Pará	Xanthic Ferralsols		+10.3		2
Brazil	Amazonas	Xanthic Ferralsols		+7.4		2
Brazil	Paraná	Rhodic Ferralsols	0–10	+10.9		2
U.S.A.	Hawaii	Volcanic tephra	Surface	−2.2		3
U.S.A.	Hawaii	Volcanic tephra	Surface	−2.0		3
U.S.A.	Hawaii	Volcanic tephra	Surface	+1.4		3
U.S.A.	Hawaii	Volcanic tephra	Surface	−1.0		3
U.S.A.	Hawaii	Volcanic tephra	Surface	−0.8		3
U.S.A.	Hawaii	Volcanic tephra	Surface	+0.2		3
U.S.A.	Hawaii	Volcanic tephra	30–40	+1.4		3
U.S.A.	Hawaii	Volcanic tephra	30 40	0.4		3
U.S.A.	Hawaii	Volcanic tephra	30–40	+6.2		3
U.S.A.	Hawaii	Volcanic tephra	30–40	+4.7		3
U.S.A.	Hawaii	Volcanic tephra	30–40	+4.6		3
U.S.A.	Hawaii	Volcanic tephra	30–40	+5.1		3
Thailand	Chachoengsao	Gray podzolic	0–15	+7.7	0.02	4
Thailand	Chantaburi	Red yellow podzolic	0–30	+7.1	0.03	4
Thailand	Rayong	Red yellow podzolic	0–10	+8.4	0.06	4
Thailand	Chantaburi	Regosols		+3.7	0.04	4
Thailand	Chainat	Gray lowland	0–10	+9.5	0.08	4
U.S.A.	Wiscosin	Typic Hapludalfs	0–10	+2.6	0.25	5
U.S.A.	Wisconsin	Typic Hapludalfs	10–20	+5.3	0.08	5
U.S.A.	Maine	Caribou	0–11.4	+4.9		6
U.S.A.	Maine	Caribou	11.4–12.7	+8.5		6

Table 2. Continued.

Country	Region	Soil type	*z*	^{15}N	%N	Re
U.S.A.	Maine	Caribou	12.7–20.3	+8.3		6
U.S.A.	Maine	Caribou	20.3–33.0	+8.0		6
U.S.A.	Maine	Caribou	33.0–43.2	+5.7		6
U.S.A.	Maine	Caribou	43.2–55.9	+3.7		6
U.S.A.	Maine	Caribou	55.9–101	+3.0		6
U.S.A.	New Hampshire			+9.6		7
U.S.A.	Central Illinois	Drummer silty clay loam		+10.4		7
U.S.A.	Oregon		0–60	+2.1		8
U.S.A.	Oregon		0–60	+2.5		8
Japan	Tokyo	Podzol	0–90	+5.0		9
Japan	Nagano	Brown Forest	0–90	+5.0		9
Belgium	Ardennes	Acid Brown	0–1	−3.9		10
Belgium	Ardennes	Acid Brown	1–4	−0.7		10
Belgium	Ardennes	Acid Brown	4–22	+0.7		10
Belgium	Ardennes	Acid Brown	22–43	+2.1		10
Belgium	Ardennes	Acid Brown	43–70	+2.6		10
Belgium	Ardennes	Acid Brown	70–100	+1.0		10
Belgium	Ardennes	Acid Brown	0–1.5	−5.1		10
Belgium	Ardennes	Acid Brown	1.5–10.5	−0.3		10
Belgium	Ardennes	Acid Brown	10–25	−0.1		10
Belgium	Ardennes	Acid Brown	25–51	+1.3		10
Belgium	Ardennes		51–80	+3.6		10
Belgium	Ardennes		80–110	+1.0		10
France	Northern Alps	Ochreous podzolic	0–1	−4.0	1.57	11
France	Northern Alps	Ochreous podzolic	2–6	−2.8	1.39	11
France	Northern Alps	Ochreous podzolic	6–11	+0.3	1.21	11
France	Northern Alps	Ochreous podzolic	11–12	+2.8	0.84	11
France	Northern Alps	Ochreous podzolic	12–14	+4.7	0.20	11
France	Northern Alps	Ochreous podzolic	14–35	+5.1	0.17	11
France	Northern Alps	Ochreous podzolic	80	+5.0	0.21	11
U.S.A.	Tennessee			+3.6		12
U.S.A.	Tennessee			+5.4		12
U.S.A.	Tennessee			+3.9		12
U.S.A.	Tennessee			+4.8		12
U.S.A.	Tennessee			+5.1		12
U.S.A.	Tennessee			+4.3		12
U.S.A.	Tennessee			+5.5		12
U.S.A.	Tennessee			+5.3		12
U.S.A.	Tennessee			−2.4		13
U.S.A.	Tennesse			−1.0		13
U.S.A.	Maine			+2.3		14
Germany	Bavaria		Olf	−3.5		15
Germany	Bavaria		Oh	+0.1		15
Germany	Bavaria		A0-5	+2.9		15

Table 2. Continued.

Country	Region	Soil type	z	15N	%N	Re
Germany	Bavaria		A5-15	+4.0		15
Sweden	Southern Sweden		Oh	−1.1		16
Sweden	Northern Sweden			−0.7		17
Sweden	Northern Sweden			+0.5		17
Sweden	Northern Sweden		Oi	−2.0		18
Sweden	Northern Sweden		Oa	−0.4		18
Sweden	Northern Sweden		Oe	+1.1		18
Sweden	Northern Sweden		A0-5	+5.0		18
U.S.A.	Massachusetts		F	+0.4		19
U.S.A.	Massachusetts		H	+4.0		19
U.S.A.	Massachusetts		10–15	+6.3		19
U.S.A.	Minnesota		Oa/A	−1.3		19
U.S.A.	Minnesota		20–25	+5.3		19

1 – Piccolo et al. (1996); 2 – Yoneyama et al. (1993) 3 – Vitousek et al. (1989); Vitousek PM (nonpublished data); 4 – Yoneyama et al. (1990); 5 – Nadelhoffer KJ and Fry B (1988); 6 – Shearer et al. (1978); 7 – Shearer et al. (1974); 8 – Binkley et al. (1992); 9 – Wada et al. (1984); 10 – Riga et al. (1971); 11 – Mariotti et al. (1980); 12 – Garten Jr & Miegroet (1994); 13 – Garten Jr CT (1993); 14 – Nadelhoffer et al. (1995); 15 – Gebauer and Schulze (1991); 16 – Nasholm et al. (1997); 17 – Michelsen et al. (1996); 18 – Hogberg et al. (1996); 19 – Fry (1991).

we expected that in forests on white sand soils, such as the Campina site, δ^{15}N values would be lower than those from the relatively N-rich Terra-firme and Varzea forests. Indeed, the only negative δ^{15}N values among all lowland tropical forests occurred in the two white sand sites: the Brazilian Campina and the Kerangas site in Sarawak. The average value of these sites was significantly lower ($P < 0.001$) than for any of the other tropical forests types we surveyed.

Within the Terra-firme forests, the site which had higher elemental concentrations in leaves (Samuel) also showed an average δ^{15}N value significantly higher ($P < 0.001$) than the Terra-firme forest at Reserva Ducke in Manaus (Table 3). In contrast, the Varzea forest had the highest elemental concentration in leaves, but a significantly lower δ^{15}N in relation to the two Terra-firme forests. Isotopic values for surface soil samples (0–10 cm) were also significantly lower in Varzea (^{15}N = +4.1±0.5‰, $n = 17$) than in Terra-firme (δ^{15}N = +9.21.5‰, $n = 10$) sites. One possible cause for the lower δ^{15}N values in the Varzea is the very high rate of nitrogen fixation by legumes (and probably by Paspalum grasses) that occurs in this system (Martinelli et al. 1992); fixed N has a δ^{15}N value close to zero. In addition, Varzea soils are formed and renewed each year by sediments brought by the white-water rivers

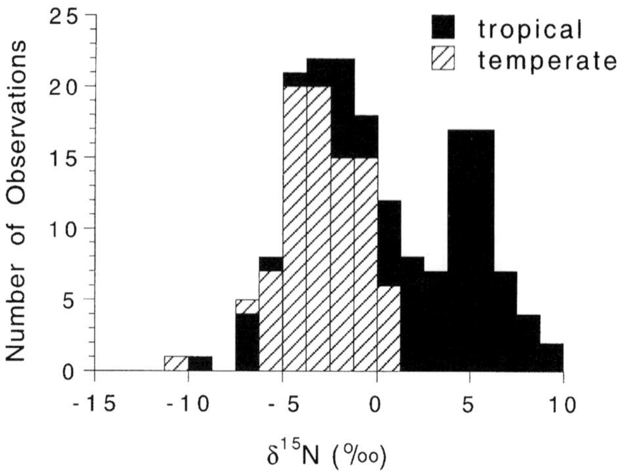

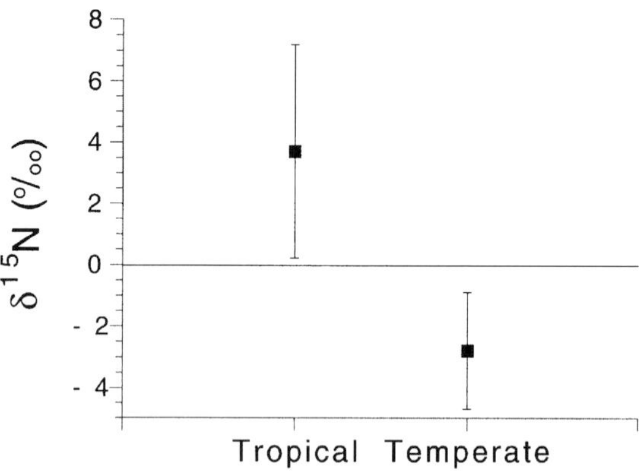

Figure 1. Histogram of $\delta^{15}N$ (‰) for tree leaves collected in tropical and temperate forests. Solid bars are tropical sites and hatched bars are temperate sites. (b) Plot of $\delta^{15}N$ (‰) for tree leaves collected in tropical and temperate forests. Error bars represent one standard-deviation.

of the Amazon, and therefore substrate age differences between Varzea and Terra-firme sites may also contribute to the differences in ^{15}N. It is possible that the Varzea site has not been in place long enough for ^{15}N enrichment to occur to the same extent as in Terra-firme. This age effect is discussed in the next section.

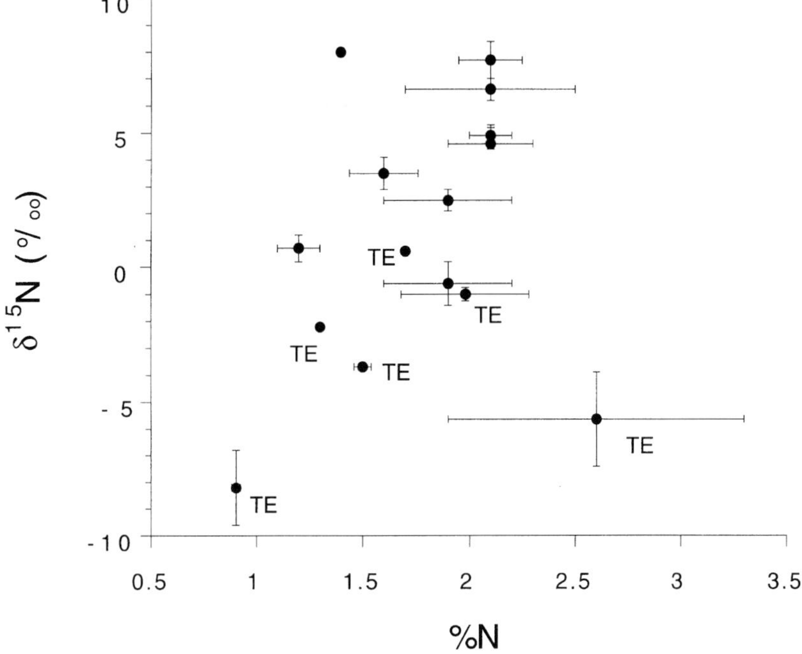

Figure 2. Relationship between average $\delta^{15}N$ values (‰) and average nitrogen concentration (%) of foliar samples. Temperate sites are labeled "TE", the unlabeled points are tropical sites. Bars are standard errors.

The Hawaii substrate age gradient and $\delta^{15}N$ of plants and soils

We used a developmental sequence of montane forests in the Hawaiian Islands to evaluate how losses of N by fractionating pathways could shape

Table 3. Average $\delta^{15}N$ value (‰) and average elemental composition of leaves (%) from distinct forest types in the Amazon Basin and averaged $\delta^{15}N$ (‰) of Kerangas site in Sarawak.

Vegetation	$\delta^{15}N$	N	P	K	Ca	Mg
Varzea[1]	+2.5	2.19	0.19	1.23	1.34	0.36
Terra-Firme[2]	+6.7	1.90	0.08	0.77	0.77	0.33
Terra-Firme[3]	+4.5	1.84	0.05	0.43	0.43	0.29
Campina[4]	−3.2	1.11	0.05	0.37	0.37	0.26

[1] Inundation forest – Senna (1996); [2] Terra-firme forest at Samuel, Rondônia – Almeida (1995); [3] Terra-firme forest at Manaus – Klinge et al. (1984); [4] White-sand soil forest (campina) near Manaus – Klinge et al. (1984).

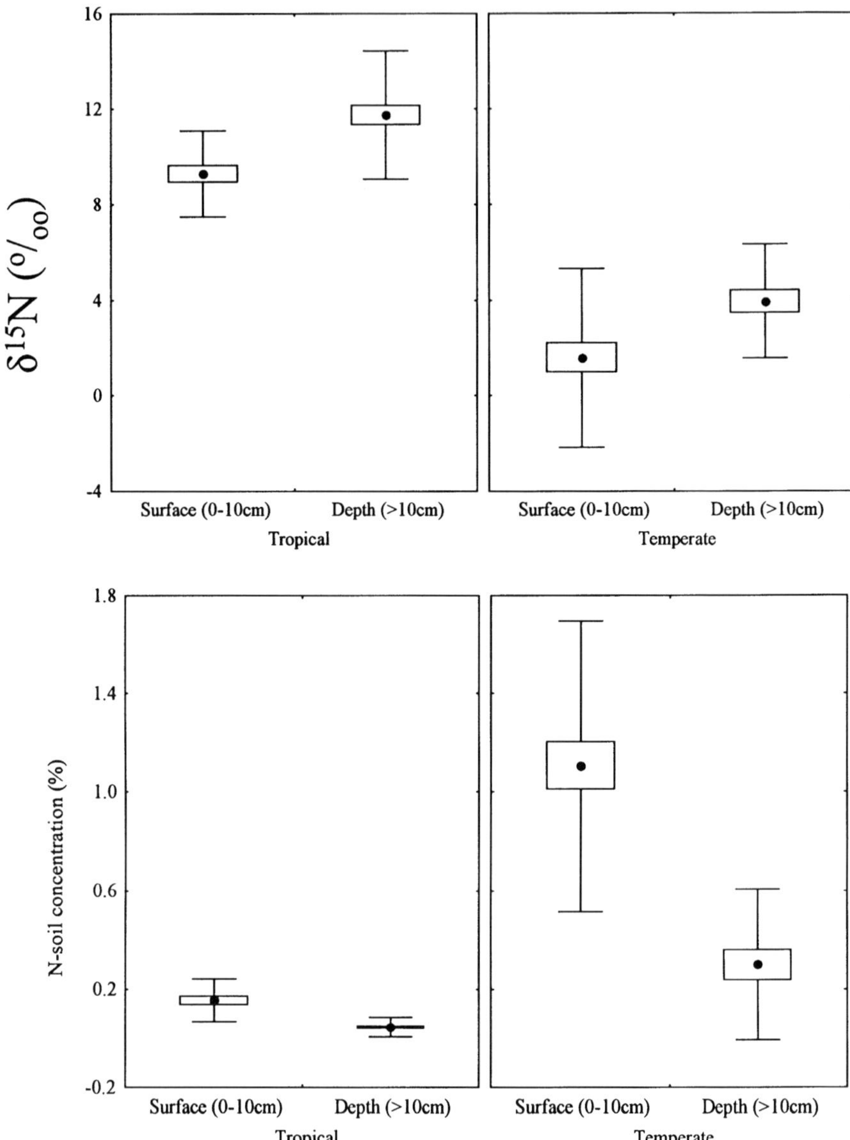

Figure 3. Box-whisker plot of (a) $\delta^{15}N$ (‰) and (b) nitrogen concentration (systems. The circles are average values, boxes are standard-errors, and bars represent one standard-deviation.

^{15}N values in forest ecosystems over time. The Hawaiian Islands result from the movement of the Pacific tectonic plate over a stationary convective plume or "hot-spot" in Earth's mantle. The hot-spot is now located under the active volcanoes at the southeast edge of the Hawaiian chain; the age of the islands increases progressively to the northwest.

Six sites with substrate ages of 300, 2100, 20,000, 150,000, 1.4 million, and 4.1 million years, located across the Islands, have been used in several soil and biogeochemical studies (Crews et al. 1995; Vitousek et al. 1997). All of the sites have basaltic parent material, are on constructional shield volcanic surfaces, at 1200 m elevation, and with 2500 mm annual precipitation. All are dominated by the native myrtaceous tree *Metrosideros polymorpha*; none have been cleared by humans. No symbiotic N fixers are important components of any site. Fertilization studies show that productivity in the youngest sites is profoundly limited by low levels of N availability (Vitousek et al. 1993); N concentrations in plant tissues, inorganic N pools, rates of soil N transformations, and gaseous losses of N are also low in these sites (Riley & Vitousek 1995; Vitousek et al. 1995). By 20,000 years, more N has accumulated and N alone no longer limits production (Vitousek & Farrington 1997). Nitrogen concentration, transformations, and trace gas losses are all much greater than in the younger sites (Vitousek et al. 1995; Crews et al. 1995). Finally, plant production in the oldest site is limited by P (Herbert & Fownes 1995), while rates of N transformations and gaseous losses remain high (Riley & Vitousek 1995). We anticipated that if losses of N by fractionating pathways drive ^{15}N enrichment, then the youngest sites should have the lowest δ^{15}N; they should be accumulating N from the atmosphere, with little N loss. As nitrification and losses of nitrate through leaching and/or denitrification increase through soil and ecosystem development, δ^{15}N in the systems as a whole should become enriched.

We measured δ^{15}N in the foliage of the dominant tree *Metrosideros polymorpha*, in the subcanopy tree fern *Cibotium glaucum*, and in surface and subsurface soils across the developmental sequence. In plants, we found the expected pattern of strongly depleted ^{15}N (to levels comparable to those in many temperate forests) in the youngest sites. Foliar δ^{15}N was enriched by 8–10‰ by the 20,000 year-old site, where foliar N concentrations and several other measures of N transformation and loss peaked. Thereafter, δ^{15}N became 1–3‰ more depleted, never approaching the very low levels observed in the youngest sites (Figure 4). δ^{15}N values in soils yielded a similar pattern, although variation in soils across the sequence was less than that in plants. All soils were enriched relative to plants, and enriched at depth relative to the surface (Figure 4).

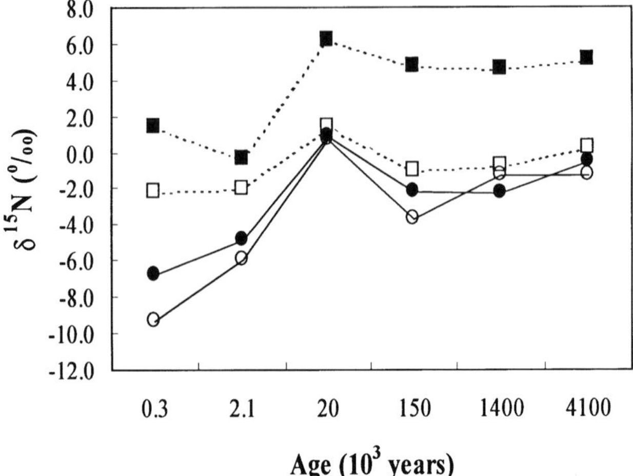

Figure 4. δ^{15}N (‰) variation for surface (open squares) and deeper (filled squares) soil samples, and for leaves of *Metrosideros polymorpha* (open circles) and *Cibotium glaucum* (filled circles) along a gradient of substrate age in Hawaii.

Overall, the observed pattern is consistent with the hypothesis that N losses by fractionating pathways cause plant and soil δ^{15}N to become relatively enriched in sites where N supply is less- or nonlimiting to biological processes. In contrast, plant and soil ^{15}N is relatively depleted in N-limited areas.

^{15}*N enrichment with soil depth*

A common feature in soil profiles is ^{15}N enrichment with soil depth. In order to compare different soil profiles, we defined the value Δ^{15}N as the difference between the δ^{15}N values of a soil depth and the δ^{15}N values of the soil surface. Table 4 summarizes the largest Δ^{15}N values found for soil profiles in tropical and temperate regions. In the tropical region, with the notable exception of the Manaus site, the Δ^{15}N values varied from 1.1 to 4.2‰, averaging 2.2±1.0‰ (n = 9). When the Δ^{15}N value of the Manaus site is included (14%), the average enrichment increases to 3.3±3.9‰ (n = 10). In the temperate sites, Δ^{15}N values varied from 2.7 to 9.1‰ with an average of 6.4±2.4‰ (n = 7). These differ significantly at the 0.2‰ level, if the Manaus site is excluded. With the Manaus value included, the Δ^{15}N is still marginally higher in the temperate zone (p = 0.06). The causes for ^{15}N enrichment with depth are relatively well known (Wada et al. 1984; Nadelhoffer & Fry 1988; Piccolo et al. 1994; Piccolo et al. 1996). More intriguing is the different behavior for soil profiles in the tropics versus temperate zones. Even among tropical soils,

Table 4. Isotopic enrichment with soil depth ($\Delta^{15}N$) for tropical and temperate regions. Depth of largest $\Delta^{15}N$ is the soil depth where the maximum isotopic enrichment occurred and Maximum depth is the deepest soil layer sampled.

Region	Site	Depth of largest $\Delta^{15}N$ (cm)	Maximum depth (cm)	$\Delta^{15}N$* (‰)
Tropical	Agua Parada	40–50	150	3.4
	Piqui	20–30	130	1.5
	Manaus	140	140	14.0
	Porto Velho	20–30	30	1.1
	Jamari	20–30	30	1.8
	Cacaulandia	20–30	30	1.2
	Vilhena-1	20–30	30	2.5
	Vilhena-2	20–30	30	4.2
	Nova Vida	20–30	130	2.4
	Benjamin	30–40	100	1.6
Temperate	Presque Isle	11–12	100	3.6
	Saint-Hubert	43–70	100	6.6
	Willerzie	51–80	110	8.6
	Chablais	14–35	80	9.1
	Bavaria	Horizon A (5–15 cm)	Horizon Oh	7.5
	Northern Sweden	Horizon A (0–5 cm)	Horizon A	7.0
	Madison	10–20	20	2.7

* $\Delta^{15}N$ is defined as $\delta^{15}N$ of depth i – $\delta^{15}N$ of surface .

the reason for the large $\Delta^{15}N$ value found in the Manaus profile is unclear; Piccolo et al. (1996) suggested that the very short dry season in the Manaus region makes losses of N higher in relation to other sites in the Amazon, where the dry season is more prolonged. However, the same explanation is unlikely when comparing tropical and temperate soils, since losses of nitrogen appear to be higher in tropical sites (Matson & Vitousek 1990; Keller et al. 1993; Neill et al. 1995). An alternative explanation could be based upon the effects of a higher cation exchange capacity and longer residence time of nitrogen in temperate soils. Higher exchange cation capacity would enhance the capability of the exchange complex to discriminate against ^{14}N, which would be preferentially lost from the profile, though the few measurements that exist suggest that ion exchange isotope effects may be minimal (B. Fry, pers. comm.). Finally, the observed differences between temperate and tropical soil profiles may be partly a function of soil age.

Conclusions

The δ^{15}N values of plants and soils are readily measured, but not always readily interpreted. We hypothesized that if N functions as an excess nutrient in many lowland tropical forests, then losses of N by fractionating pathways would cause δ^{15}N to become enriched in those forests. The results of our survey confirm this prediction: tropical forests in general are ^{15}N-enriched compared to temperate forests, and tropical forests in which N is abundant are ^{15}N-enriched compared to tropical forests in which N appears to be limiting. These ^{15}N results do not establish unequivocally that N is in excess in tropical forests; mechanisms other than N loss by fractionating pathways could cause part of the observed signal. For example, if N fixation is relatively more important than atmospheric deposition in tropical forests, that could account for part – although not all – of the observed difference. However, the pattern of ^{15}N enrichment in tropical forests does provide an additional, time integrated line of evidence supporting the conclusion that many tropical forests are N-rich and have open N cycles in comparison to most temperate forests.

Acknowledgements

We would like to thank Michael McClain for providing leaf samples from the Campina site, and Brian Fry for a critical review that improved an earlier draft of this manuscript.

References

Aber JD, Magill A, McNulty SG, Boone RD, Nadelhoffer RJ, Downs M & Hallett R (1995) Forest biogeochemistry and primary production altered by nitrogen saturation. Water Air and Soil Pollut. 85: 1665–1670

Almeida S (1995) Dinâmica de nutrientes e variação natural de ^{13}C e ^{15}N de uma floresta tropical Húmida de terra-firme, Estação Ecológica de Samuel, RO (BR). PhD Thesis. University of São Paulo, Piracicaba, Brazil

Berendse F, Aerts R & Bobbink R (1993) Atmospheric nitrogen deposition and its impact on terrestrial ecosystems. In: Vos CC & Opdam P (Ed.) Landscape Ecology of a Stressed Environment (pp 104–121). Chapman & Hall, London, England

Binkley D, Sollins P, Bell R, Sachs D & Myrold D (1992) Biogeochemistry of adjacent alder and alder-conifer stands. Ecology 73: 2022–2033

Crews TE, Kitayama K, Fownes J, Herbert D, Mueller-Dombois D, Riley RH & Vitousek PM (1995) Changes in soil phosphorus and ecosystem dynamics across a long soil chronosequence in Hawaii. Ecology 76: 1407–1424

Cuevas E & Medina E (1988) Nutrient dynamics within Amazonian forests. II fine root growth, nutrient availability and leaf litter decomposition. Oecologia 76: 222–235

Domenach AM, Kurdali F & Bardin R (1989) Estimation of symbiotic dinitrogen fixation in alder forest by the method based on natural ^{15}N abundance. Plant Soil 118: 51–59

Fry B (1991) Stable isotope diagrams of freshwater food webs. Ecology 72: 2293–2297

Furch K & Klinge H (1989) Chemical relationships between vegetation, soil and water in contrasting inundation areas of Amazonia. In: Proctor J (Ed.) Mineral Nutrients in Tropical Forest and Savanna Ecosystems (pp 189–204). Blackwell Scientific Publications, Oxford

Garten Jr CT (1993) Variation in foliar ^{15}N abundance and the availability of soil nitrogen on Walker Branch watershed. Ecology 74: 2098–2113

Garten Jr CT & Miegroet HV (1994) Relationships between soil nitrogen dynamics and natural ^{15}N abundance in plant foliage from Great Smoky Mountains National Park. Can. J. Forest. Res. 24: 1636–1645

Gebauer G & Schulze ED (1991) Carbon and nitrogen isotope ratios in different compartments of a healthy and a declining Picea abies forest in the Fitchtelgebirge, NE Bavaria. Oecologia 87: 198–207

Hedin LO, Armesto JJ & Johnson AH (1995) Patterns of nutrient loss from unpolluted, old-growth temperate forests: Evaluation of biogeochemical theory. Ecology 76: 493–509

Herbert DA & Fownes JH (1995) Phosphorus limitation of forest leaf area and net primary productivity on highly weathered tropical montane soils in Hawaii. Biogeochem. 29: 223–235

Hill PW, Handley LL & Raven JA (1996) Juniperus communis L.ssp. at Balnaguard, Scotland: foliar carbon discrimination (^{13}C) and ^{15}N natural abundance (^{15}N) suggest gender-linked differences in water and N use. Bot. J. Scotl. 48(2): 209–224

Hogberg P & Alexander IJ (1995) Roles of root symbioses in African woodland and forest: Evidence from ^{15}N abundance and foliar analysis. J. Ecology 83: 217–224

Hogberg P, Hogbom L, Schinkel H, Hogberg M, Johannisson C & Wallmark H (1996) ^{15}N abundance of surface soils, roots and mycorrhizas in profiles of European forest soils. Oecologia 108: 207–214

Hogberg P (1997) ^{15}N natural abundance in soil-plant systems. New Phytologist 137: 179–203

Keller M, Kaplan WA & Wofsy SC (1986) Emissions of N_2O, CH_4, and CO_2 from tropical soils. J. Geophys. Res. 91: 11791–11802

Keller M, Veldkamp E, Weitz AM & Reiners WA (1993) Effect of pasture age on soil trace-gas emissions from a deforested area of Costa Rica. Nature 365: 244–246

Klinge H, Furch K & Harms E (1984) Selected bioelements in bark and wood of native tree species from Central-Amazonian inundation forests. Amazoniana 9(1): 105–117

Lewis WM (1986) Nitrogen and phosphorus runoff losses from a nutrient-poor tropical moist forest. Ecology 67: 1275–1282

Mariotti A, Pierre D, Vedy JC, Bruckert S & Guillemot J (1980) The abundance of natural Nitrogen 15 in the organic matter of soils along an altitudinal gradient (Chablais, Haute Savoie, France). Catena 7: 293–300

Martinelli LA, Victoria RL, Trivelin PCO, Devol AH & Richey JE (1992) ^{15}N natural abundance in plants of the Amazon river floodplain and potential atmospheric N_2 fixation. Oecologia 90: 591–596

Matson PA, Vitousek PM, Ewel JJ, Mazzarino MJ & Robertson GP (1987) Nitrogen transformations following tropical forest felling and burning on a volcanic soil. Ecology 68: 491–502

Matson PA & Vitousek PM (1987) Cross-system comparison of soil nitrogen transformations and nitrous oxide fluxes in tropical forests. Global Biogeochem. Cycles 1: 163–170

Matson PA & Vitousek PM (1990) Ecosystem approach to a global nitrous oxide budget. BioSci. 40: 667–672

Matson PA, McDowell WH, Townsend AR & Vitousek PM (1999) The globalization of N deposition: Ecosystem consequences in tropical environments.

Michelsen A, Schmidt IK, Jonasson S, Quarmby C & Sleep D (1996) Leaf ^{15}N abundance of subarctic plants provides field evidence that ericoid, ectomycorrhizal and non- and arbuscular mycorrhizal species access different sources of soil nitrogen. Oecologia 105: 53–63

Nadelhoffer KJ & Fry B (1988) Controls on natural Nitrogen-15 and Carbon-13 abundances in forest soil organic matter. Soil Sci. Soc. of Am. J. 52: 1633–1640

Nadelhoffer KJ, Downs MR, Fry B, Aber JD, Magill AH & Melillo JM (1995) The fate of ^{15}N-labelled nitrate additions to a northern hardwood forest in eastern Maine (USA). Oecologia 103: 292–301

Nasholm T, Nordin A, Edfast AB & Hogberg P (1997) Identification of coniferous forests with incipient nitrogen saturation through analysis of arginine and nitrogen-15 abundance of trees. J. Environ. Qual. 26: 302–309

Neill C, Piccolo MC, Steudler PA, Melillo JM, Feigl BJ & Cerri CC (1995) Nitrogen dynamics in soils of forests and active pastures in the western Brazilian Amazon Basin. Soil Biol. Biochem. 27: 1167–1175

Piccolo MC, Neill C & Cerri CC (1994) ^{15}N natural abundance in soils along forest-to-pasture chronosequences in the western Brazilian Amazon Basin. Oecologia 99: 112–117

Piccolo MC, Neil C, Mellilo JM, Cerri CC & Steudler PA (1996) ^{15}N natural abundance in forest and pasture soils of the Brazilian Amazon Basin. Plant Soil 182: 249–258

Proctor J, Anderson JM & Vallack HW (1983) Comparative studies on soils and litterfall in forests at a range of altitudes on Gunung Mulu, Sarawak. Malaysian Forester 46: 60–76

Riga A, Van Praag HJ & Brigodi N (1971) Rapport isotopique natural de l'azote dans quelques sols forestriers et agricoles de Belgique soumis'a difeves traitments culturaux. Geoderma 6: 213–222

Riley RH & Vitousek PM (1995) Nutrient dynamics and trace gas flux during ecosystem development in Hawaiian montane rainforest. Ecology 76: 292–304

Salati E, Sylvester-Bradley R & Victoria RL (1982) Regional gains and losses of nitrogen in the Amazon Basin. Plant Soil 67: 367–376

Senna HBC (1996) Avaliação dos teores de nitrogênio, fósforo, potássio, cálcio, magnésio e variação natural do ^{13}C em florestas de inundação na Amazônia Central. PhD Thesis. University of São Paulo, Piracicaba, Brazil

Schulze ED, Chapin III FS & Gebauer G (1994) Nitrogen nutrition and isotope differences among life forms at the northern treeline of Alaska. Oecologia 100: 406–412

Shearer G, Kohl DH, & Chien SH (1978) The Nitrogen-15 abundance in a wide variety of soils. Soil Sci. Soc. Am. J. 42: 899–902

Shearer G, Duffy J, Kohl DH & Commoner BA (1974) Steady-state model of isotopic fractionation accompanying nitrogen transformations in soil. Soil Sci. Soc. Am. J. 38: 315–322

Sprent JI, Geoghegan IE, Whitty PW & James EK (1996) Natural abundance of ^{15}N and ^{13}C in nodulated legumes and other plants in the cerrado and neighbouring regions of Brazil. Oecologia 105: 440–446

Tanner EVJ, Vitousek PM & Cuevas E (1998) Experimental investigation of the role of nutrient supplies in the limitation of forest growth and stature on wet tropical mountains. Ecology 79: 10–22

Virginia RA & Delwiche CC (1982) Natural 15N abundance of presumed N_2-fixing and non-N_2-fixing plants from selected ecosystems. Oecologia 54: 317–325

Vitousek PM (1984) Litterfall, nutrient cycling, and nutrient limitation in tropical forests. Ecology 65: 285–298

Vitousek PM & Sanford RL (1986) Nutrient Cycling in Moist Tropical Forest. Annu. Rev. Ecol. Syst. 17: 137–167

Vitousek PM & Denslow JS (1986) Nitrogen and phosphorus availability in treefall gaps in a lowland tropical rainforest. J. Ecology 74: 1167–1178

Vitousek P, Shearer G & Kohl DH (1989) Foliar ^{15}N natural abundance in Hawaiian rainforest: Patterns and possible mechanisms. Oecologia 78: 383–388

Vitousek PM, Walker LR, Whiteaker LD & Matson PA (1993) Nutrient limitation to plant growth during primary succcession in Hawaii Volcanoes National Park. Biogeochem. 23: 197–215

Vitousek PM, Turner DR & Kitayama K (1995) Foliar nutrients during long-term soil development in Hawaiian montane rain forest. Ecology 76: 712–720

Vitousek PM, Chadwick OA, Crews T, Fownes J, Hendricks D & Herbert D (1997) Soil and ecosystem development across the Hawaiian Islands. GSA Today 7: 1–8

Vitousek PM & Farrington H (1997) Nutrient limitation and soil development: Experimental test of a biogeochemical theory. Biogeochem. 37: 63–75

Vogt KA, Grier CC & Vogt DJ (1986) Production, turnover, and nutrient dynamics of above and below-ground detritus of world forests. Advances in Ecological Research 15: 303–377

Yoneyama T, Murakami T, Boonkerd N, Wadisirisuk P, Siripin S & Kouno K. (1990) Natural ^{15}N abundance in shrub and tree legumes, *Casuarina*, and non N_2 fixing plants in Thiland. Plant Soil 128: 287–292

Yoneyama T, Muraoka T, Murakami T, & Boonkerd N (1993) Natural abundance of ^{15}N in tropical plants with emphasis on tree legumes. Plant Soil 153: 295–304

Wada E, Imaizumi R & Takai Y (1984) Natural abundance of ^{15}N in soil organic matter with special reference to paddy soils in Japan: Biogeochemical implications on the nitrogen cycle. Geochem. J. 18: 109–123

Biogeochemistry **46:** 67–83, 1999.

The globalization of N deposition: ecosystem consequences in tropical environments

PAMELA A. MATSON[1], WILLIAM H. McDOWELL[2], ALAN R. TOWNSEND[3] & PETER M. VITOUSEK[4]

[1]*Department of Geological and Environmental Sciences and the Institute for International Studies, Stanford University, Stanford, CA 94305-2115 U.S.A.;* [2]*Department of Natural Resources, University of New Hampshire, Durham, NH 03824 U.S.A.;* [3]*Department of Evolutionary, Population and Organismic Biology and the Institute for Arctic and Alpine Research, University of Colorado, Boulder, CO 80303 U.S.A.;* [4]*Department of Biological Sciences, Stanford University, Stanford, CA 94305 U.S.A.*

Received 10 December 1998

Key words: acidification, anthropogenic nitrogen, cations, nitrate leaching, nitric oxide, nitrous oxide, nutrient limitation, phosphorus, productivity, tropical ecosystems

Abstract. Human activities have more than doubled the inputs of nitrogen (N) into terrestrial systems globally. The sources and distribution of anthropogenic N, including N fertilization and N fixed during fossil fuel combustion, are rapidly shifting from the temperate zone to a more global distribution. The consequences of anthropogenic N deposition for ecosystem processes and N losses have been studied primarily in N-limited ecosystems in the temperate zone; there is reason to expect that tropical ecosystems, where plant growth is most often limited by some other resource, will respond differently to increasing deposition. In this paper, we assess the likely direct and indirect effects of increasing anthropogenic N inputs on tropical ecosytem processes. We conclude that anthropogenic inputs of N into tropical forests are unlikely to increase productivity and may even decrease it due to indirect effects on acidity and the availability of phosphorus and cations. We also suggest that the direct effects of anthropogenic N deposition on N cycling processes will lead to increased fluxes at the soil-water and soil-air interfaces, with little or no lag in response time. Finally, we discuss the uncertainties inherent in this analysis, and outline future research that is needed to address those uncertainties.

Introduction

Human activity has more than doubled the quantity of nitrogen (N) fixed in terrestrial ecosystems, due to industrial N fixation, the mobilization and inadvertent fixation of nitrogen during fossil fuel combustion, and cultivation of nitrogen fixing crops (Smil 1990; Galloway et al. 1995; Vitousek et al. 1997). The mobility of fixed N within and between terrestrial ecosystems has also been enhanced as a consequence of land clearing, biomass burning, additions of anthropogenic N are likely to have little direct effect on carbon uptake in most tropical forests.

Nitrogen biogeochemical processes in soils strongly reflect the greater availability of N in many tropical soils. Studies of N cycling and N trace gas emissions across a range of tropical ecosystems suggest that N gas fluxes (as a proportion of N mineralization) are greater in tropical than in temperate systems, and that quantitatively, N trace gas fluxes are on average much greater in tropical than in temperate systems (Keller et al. 1986; Matson & Vitousek 1987; Matson et al. 1990; Hall et al. 1996). While Vitousek and Matson (1988) reported a range in microbial uptake of applied ^{15}N depending on tropical soil type, microbial immobilization of ^{15}N generally was much less than in temperate systems where similar measurements have been made (e.g., Matson et al. 1992). Thus, it seems likely that retention of anthropogenic N will be much less in most tropical systems than in most temperate systems, due to reduced microbial uptake and retention in soil organic matter as well as reduced uptake and retention in plant biomass.

The rather limited data from upland tropical forests also suggests that most mineralized N is quickly nitrified (Vitousek & Matson 1988; McDowell et al. 1992, 1996). In contrast, in many temperate forests, not all N that is mineralized to ammonium is oxidized to nitrate (Vitousek et al. 1982). Indeed, one of the hypothesized responses of temperate soils to long-term anthropogenic N inputs is a gradual shift in the nitrogen economy of soils, from ammonium-dominated to nitrate-dominated systems (Melillo et al. 1989; Aber 1992). Upland tropical soils are likely to display little delay in nitrification in response to increasing N additions.

Indirect effects of anthropogenic N on ecosystem processes

The biogeochemical consequences of enhanced N inputs to terrestrial ecosystems can range well beyond direct alterations of nitrogen cycling and plant response. Numerous studies in natural and agricultural systems of the temperate zone have shown that both atmospheric and fertilizer inputs of N can lead to soil acidification, depletion of base cations, and mobilization of potentially toxic aluminum (Al) ions (Johnson et al. 1982; Hallbacken & Tamm 1986; Aber et al. 1989; Schulze 1989; Lawrence et al. 1995; Likens et al. 1996). Moreover, many tropical soils have developed in place over a very long time, without rejuvenation by glaciation or loess deposition. Accordingly, these soils can be highly weathered, depleted in primary minerals, poorly buffered, and dominated by variable charge clays (Uehara & Gillman 1981; Sanchez & Logan 1992). These systems are therefore (1) poorly buffered against additional sources of acidity, and (2) potentially quite sensitive to base cation depletion. As N inputs increase in humid tropical regions, changes in soil acidity, base cation supply, P availability, Al mobility, and carbon storage are all possible, but the rates, consequences, and sometimes even the direction of such changes are likely to be markedly different in many tropical sites than what has been observed in the temperate zone.

Soil acidification

Elevated N inputs can lead to soil acidification via several pathways, with the overall change depending on the properties of the ecosystem, the form in which N is added, and the anion or cation associated with the added N (Ruess & Johnson 1986). Both fertilizer and atmospheric N inputs can lead to soil acidification; our focus here will be upon atmospheric inputs only. Added N can be acidifying regardless of the inorganic form by which it is added, but the net change in soil acidity due to inputs depends on plant uptake (Figure 1). For example, nitric acid inputs do not necessarily lead to soil acidification: if the nitrate ion is taken up and retained by plants, the proton is neutralized by release of a hydroxyl ion. However, if NO_3^- remains mobile and leaches out of the system, the proton's greater affinity for cation exchange sites in the soil will displace a base cation (potassium, calcium, magnesium, sodium) which will balance the charge of the leaching NO_3^- ion. The addition of one molecule of nitric acid would then cause a net increase of one proton in the soil and the net loss of one base cation. Addition of N as ammonium (NH_4^+) can produce the greatest per molecule effects. If the NH_4^+ ion is taken up, one proton is released. (When soil NH_4^+ is derived from decomposition of organic matter, this released proton is not an increase in acidity as it balances the consumption of a proton during conversion of NH_3 to NH_4^+.) If the added NH_4^+ ion is nitrified, two protons are released. If the resulting NO_3^- ion is then taken up by a plant, the net change is still an increase of one proton; if the NO_3^- is leached, two protons are added and one base cation is lost.

In temperate ecosystems, addition of excess N from the atmosphere has led to soil acidification and base cation depletion, but strong plant demand for N slows the rate of change. As discussed above, however, moist tropical systems frequently are rich in N relative to other essential elements. Thus, the majority of excess N inputs are likely to end up as NO_3^-, and much of this NO_3^- may not be taken up by plants. Leaching of NO_3^- from surface soils mobilizes a cation (Figure 1); as the base cation supply is progressively depleted, the leached cation will either be a proton or mobilized aluminum ion, both of which have potentially serious consequences for downstream aquatic systems (see below).

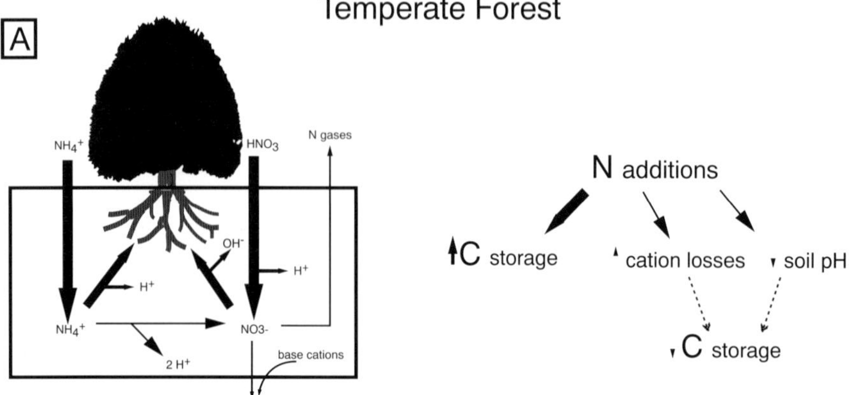

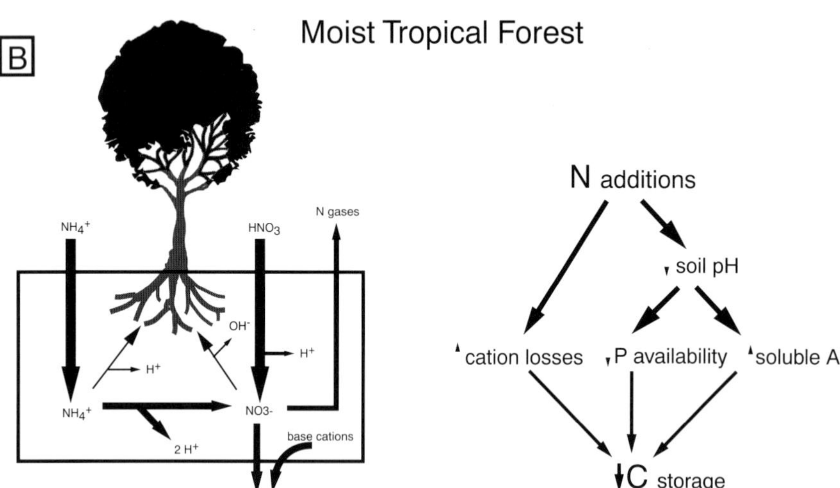

Figure 1. Diagram comparing initial responses of temperate (A) vs moist tropical (B) forests to elevated inputs of atmospheric nitrogen. The figure assumes that plant growth and carbon storage in the majority of temperate forests are initially limited by N supply, and that N functions in relative excess in the majority of lowland moist tropical forests. The thickness of the arrows represents the relative magnitude of fluxes of N in the diagrams on the left, and the relative magnitude of the effects of those fluxes in the summary flow charts on the right. Thus, in (A), we contend that most of the N additions will be retained in the system and initially lead to C storage, whereas in (B), we argue that most of the additional N inputs will be lost from the system, and that the consequences of increased nitrification rates and N losses will be losses of base cations and decreases in soil pH, which may in turn lead to decreases in C storage in moist tropical forests.

Anion adsorption capacity

The fate of excess N and the rate of soil acidification in tropical systems may also be affected by the surface charge properties of many tropical soils. Soil surface charge in the vast majority of temperate zone soils is both permanent and negative, so that anion adsorption capacity is usually quite low (Johnson & Cole 1980). However, iron (Fe) and Al oxide rich tropical soils are often of variable charge, meaning that overall net charge is a function of pH and soil solution chemistry. At the low pH's typical of these soils, net charge is frequently positive (Uehara & Gilman 1981). In addition, free Fe and Al oxides create significant sources of positive charge across a large range of pH values (Ji 1997). Thus, electrostatic adsorption of may be important in controlling losses of excess NO_3^- to aquatic systems.

Such adsorption may be especially important in deeper soil horizons. High concentrations of organic matter in surface horizons create negatively charged surfaces that make nitrate highly mobile. However, SOM decreases with depth while the density of positively charged sites increases. Matson et al. (1987) showed that significant amounts of N lost from surface horizons in a Costa Rican soil were retained as exchangeable NO_3^- in soil layers below 40cm. Soils in the lowland tropics frequently are very deep (>10 m, at times much greater – Nepstad et al. 1994; Richter & Markewitz 1995), thereby creating a long hydrologic path over which NO_3^- can be adsorbed. Thus, increases in NO_3^- leaching from surface soils due to greater N deposition in the tropics may not lead to similar increases in NO_3^- loading of aquatic ecosystems.

Prediction of the role of anion exchange in controlling NO_3^- losses is further complicated by the biogeochemistry of other elements. Where excess S inputs occur in addition to increasing N, ligand exchange of sulfate ions will decrease NO_3^- adsorption capacities (Zhang & Yu 1997). In addition, there is evidence that NO_3^- adsorption may vary considerably with the identity of its charge-balancing cation in soil solution. For example, Wang et al. (1987) found that NO_3^- adsorption in both a Brazilian and a Chinese oxisol was 15–20% greater when the accompanying cation was calcium as opposed to potassium. Given that soil acidification should result in base cation losses, NO_3^- retention may vary with time as the balance of cations in soil solution shifts.

Effects on carbon storage

Unlike the temperate zone, where increasing N deposition may cause at least a transient increase in carbon storage, we suggest that higher N inputs to moist tropical systems may lead to lower productivity and *reduced* carbon storage. This may occur via several mechanisms. First, because plant growth in many tropical systems appears limited by some combination of P and/or base cation availability, losses of base cations due to increased leaching of nitrate may cause reductions in plant growth and carbon storage. Furthermore, cation exchange capacity (CEC) in many tropical soils is largely a function of soil organic matter content (Tiessen et al. 1994a; Mogollon & Querales 1995). Thus, increasing N inputs could induce a positive feedback in which higher losses of cations decrease growth rates and organic matter pools, which in turn reduce the soil's capacity to retain base cations, further reducing productivity and C storage.[1]

Second, decreases in productivity and carbon storage may also occur due to the effects of increasing soil acidity on phosphorus availability. Phosphorus is the element most commonly associated with nutrient limitation in the lowland tropics (Sanchez et al. 1982). This deficiency occurs for two reasons. First, as with the base cations, the combination of extremely old soils and high weathering rates has led to severely depleted primary mineral pools of P (Tiessen et al. 1994b; Crews et al. 1995). Second, acidic soils rich in iron and aluminum oxides react with labile inorganic P, and fix some of that P into insoluble forms (Uehara & Gillman 1981). As a consequence, even tropical soils that have large total soil P pools may have scarce plant-available P. Additions of excess N to such soils may exacerbate P limitation, because rates of P fixation tend to increase with decreasing soil pH[2] (Edwards 1991; Pardo et al. 1992). In many tropical systems, lowered P availability is likely to result in less C uptake and storage.

Finally, the prevalence of aluminum oxides and low pH in many tropical soils means that even slight additional decreases in pH can lead to significant increases in the release of mobile Al ions into soil solution (Sanchez & Logan 1992). The solubility of Al in soils is a nonlinear function of pH, and many tropical soils have pH values well below 5, and at times below 4. Sharp increases in soluble Al occur at soil pH's below 4, thus Al mobilization is likely to occur following a much smaller increase in N deposition in the tropics than has been seen in temperate systems. Increases in soluble Al have a number of potentially deleterious effects, ranging from inhibition of plant and microbial activity to poisoning of fish and other aquatic organisms in downstream systems (NRC 1986; Ruess & Johnson 1986; Godbold et al. 1988; Shortle & Smith 1988). The inhibition of plant and microbial growth provides a third potential way in which elevated N deposition in the moist tropics may lead to net losses of carbon to the atmosphere (Figure 1).

Off site consequences

Land-water exchange

Increased N deposition in tropical landscapes is likely to have significant consequences for tropical aquatic and marine ecosystems. Although net primary productivity in most temperate lakes and rivers is P-limited, nutrient limitation appears to be less clear-cut in the tropics, where a significant number of lakes may be N-limited (Payne 1986; Vitousek & Howarth 1991) and some streams may be light-limited (Pringle et al. 1986). Near-shore marine embayments appear to be more commonly P-limited in the tropics than they are in the temperate zone, but off-shore waters are N-limited as they are in the temperate zone (Vitousek & Howarth 1991). Overall, it appears that the susceptibility of tropical waters to increased N run-off relative to their temperate zone counterparts is higher for fresh waters, less for near-shore coastal waters, and similar for off-shore ecosystems.

The extent to which increased N deposition in tropical ecosystems will affect water quality depends in part on N cycling at the land-water interface. Numerous studies in the temperate zone have shown that plant uptake and denitrification in the riparian zone typically remove 85% of dissolved N from shallow groundwater and thus protect water quality (reviewed by Triska et al. 1993). Maintenance of a riparian buffer strip in forestry and agricultural operations is considered to be 'best management practice' in many areas of the United States, as well as Europe (Lowrance et al. 1984; Vought et al. 1994).

Although fewer in number, studies of riparian N dynamics in tropical forests clearly show that N concentrations are reduced as water moves from upslope forests to streams (McDowell et al. 1992, 1996; McClain et al. 1994). In tropical forests, denitrification is thought to be the major mechanism driving this attenuation of N transfers (Bowden et al. 1992; Brandes et al. 1996). Although concentrations of NH_4^+ and DON also decline across tropical riparian zones (McDowell et al. 1996), denitrification is the process most likely to result in net loss of N from the terrestrial ecosystem to the atmosphere; thus, this pathway is most likely to provide effective long-term protection of water quality. The improvement in water quality is counterbalanced by the undesirable input of N_2O to the atmosphere, unless denitrification proceeds all the way to N_2.

The consequences of increased N loading for land-water interactions will depend in large part on the form of nitrogen lost from tropical forest soils, and on any changes in forest C dynamics that accompany increased N loading. For denitrification to proceed, anoxic conditions, a reduced carbon substrate, and nitrate must all be present. Declines in productivity and C availability

evaluating the consequences of N deposition in tropical ecosystems are not even represented in widely used ecosystem models, and all of these models are built on the concept of N limitation, not over-abundance. Only Century (Parton et al. 1988, 1995) now simulates P cycling and availability – and there is evidence that it does not work very well for tropical soils (Glijsman et al. 1996). None of the widely used models (e.g. TEM and BIOME-BGC as well as Century) evaluate cation fluxes or potential limitation. These models are crucial tools for understanding the consequences of environmental change (Schimel et al. 1997), but they require extensive modification before they can be applied to analyzing N deposition in tropical forest ecosystems.

Acknowledgements

We thank SCOPE, the Andrew Mellon Foundation, and the Inter-American Institute for Global Change for supporting the meeting that led to this special issue. Support for the authors and work contributing to this manuscript came from NSF (including LTER funding to the Luquillo Experimental Forest), USDA, the Mellon Foundation, and NASA's New Investigator Program. Comments from Alison Magill, Whendee Silver and Greg Asner substantially improved an earlier version of the paper.

Notes

1. This feedback could be counter-balanced by feedbacks to decomposition, in which reduced availability of limiting nutrients (base cations in this case) leads to lower decomposition rates and greater nutrient conservation (e.g. Vitousek 1982).
2. Although the overall effects of decreasing pH on P availability in variable charge soils are not clear – there is also some suggestion that lowering pH could cause greater anion retention, which could increase the amount of exchangeable P.

References

Aber JD (1992) Nitrogen cycling and nitrogen saturation in temperate forest ecosystems. Trends in Ecology and Evolution 7: 220–223

Aber JD, Magill A, McNulty SG, Boone RD, Nadelhoffer KJ, Downs M & Hallett R (1995) Forest biogeochemistry and primary production altered by nitrogen saturation. Water, Air and Soil Pollution 85: 1665–1670

Aber JD, Magill A, Boone RD, Melillo JM, Steudler P & Bowden R (1993) Plant and soil responses to chronic nitrogen additions at the Harvard Forest, Massachusetts. Ecological Applications 3: 156–166

Aber JD, Nadelhoffer KJ, Steudler P & Melillo JM (1989) Nitrogen saturation in northern forest ecosystems. BioScience 39: 378–386

Agren GI & Bosatta E (1988) Nitrogen saturation of terrestrial ecosystems. Environmental Pollution 54: 185–197

Andreae MO (1993) The influence of tropical biomass burning on climate and the atmospheric environment. In: Oremland RS (Ed.) Biogeochemistry of Global Change: Radiatively Active Trace Gases (pp 113–150). Chapman and Hall, NY, U.S.A.

Bowden RD, Melillo JM, Steudler PA & Aber JD (1991) Effects of nitrogen addition on annual nitrous oxide fluxes from temperate forest soils in the northeastern United States. J. Geophys. Res. 96: 9321–9328

Bowden WB, McDowell WH, Asbury CE & Finley AM (1992) Riparian nitrogen dynamics in two geomorphologically distinct tropical rain forest watersheds – nitrous oxide fluxes. Biogeochemistry 18: 77–99

Brandes JA, McClain ME & Pimentel TP (1996) ^{15}N evidence for the origin and cycling of inorganic nitrogen in a small Amazonian catchment. Biogeochemistry 34: 45–56

Crews TE, Kitayama K & Vitousek PM (1995) Changes in soil phosphorus fractions and ecosystem dynamics across a long chronosequence in Hawaii. Ecology 76: 1407–1420

Cuevas E & Medina E (1988) Nutrient dynamics within Amazonian forests: II. Fine root growth, nutrient availability, and leaf litter decomposition. Oecologia 76: 222–235

Davidson EA & Kingerlee W (1997) Global inventory of emissions of nitric oxide from soils: a literature review. Nutrient Cycling in Agroecosystems 48: 37–50

Edwards AC (1991) Soil acidity and its interactions with phosphorus availability for arange of different crop types. In: Wright RJ et al. (Eds) Plant-Soil Interactions at Low pH. Kluwer Academic Publishers, Dordrecht, The Netherlands

Eisner MJ (1990) Nitrous oxide emissions from fertilized soils: Summary of available data. J. Environ. Quality 19: 272–280

EPA 1990. Greenhouse gas emissions from agricultural ecosystems, IPCC Report. Washington, DC, U.S.A.

Galloway JN, Levy H II & Kasibhatla PS (1994) Year 2020: Consequences of population growth and development on the deposition of oxidized nitrogen. Ambio 23: 120–123

Galloway JN, Schlesinger WH, Levy H II, Michaels A & Schnoor JL (1995) Nitrogen fixation: atmospheric enhancement-environmental response. Global Biogeochemical Cycles 9: 235–252

Glijsman AJ, Oberson A, Tiessen H & Friesen DK (1996) Limited applicability of the Century model to highly weathered tropical soils. Agron. J. 88: 894–903

Godbold DL, Fritz E & Hutterman A (1988) Aluminum toxicity and forest decline. Proceedings of the National Academy of Sciences 85: 3888–3892

Grennfelt P & Hultberg H (1986) Effects of nitrogen deposition on the acidification of terrestrial and aquatic ecosystems. Water, Air, and Soil Pollution 30: 945–963

Hall SJ, Matson PA & Roth P (1996) NO_x emission from soil: Implications for air quality modeling in agricultural regions. Annual Review of Energy and Environment 21: 311–346

Hallbacken L & Tamm CO (1986) Changes in soil acidity from 1927 to 1982–1984 in a forest area of south-west Sweden. Scand. J. For. Res. 1: 219–232

Herbert DA & Fownes JH (1995) Phosphorus limitation of forest leaf area and net primary productivity on highly weathered soils. Biogeochemistry 29: 223–235

Howarth RW, Billen G, Swaney D, Townsend A, Jaworski N, Lajtha K, Downing JA, Elmgren R, Caraco N, Jordan T, Berendse F, Freney J, Kuderyarov V, Murdoch P & Zhaoliang Z (1996) Regional nitrogen budgets and riverine N and P fluxes for the drainages of the North Atlantic Ocean: Natural and human influences. Biogeochemistry 35(1): 75–140

Ji GL (1997) Electrostatic adsorption of anions. In: Yu TR (Ed) Chemistry of Variable Charge Soils (pp 113–139). Oxford University Press, New York, U.S.A.

Johnson DW & Cole DW (1980) Anion mobility in soils: Relevance to nutrient transport from forest ecosystems. Environment International 3: 79–90

Johnson DW, Turner J & Kelly JM (1982) The effects of acid precipitation on forest nutrient status. Water Resources Research 18: 449–461

Jordan CF (1985) Nutrient Cycling in Tropical Forest Ecosystems. Wiley and Sons, Chichester

Keller M & Matson PA (1994) Biosphere-atmosphere exchange of trace gases in the tropics: Evaluating the effects of land use change. In: Prinn R (Ed.) Global Atmospheric-Biospheric Chemistry. Plenum Press, NY, U.S.A.

Keller M, Kaplan WA & Wofsy SC (1986) Emissions of N_2O, CH_4, and CO_2 from tropical soils. J. Geophys. Res. 91: 11791–11802

Lawrence GB, David MB & Shortle WC (1995) A new mechanism for calcium loss in forest-floor soils. Nature 378: 162–165

Likens GE, Driscoll CT & Buso BC (1996) Long-term effects of acid rain: response and recovery of a forest ecosystem. Science 272: 244–246

Lowrance R, Todd R, Fail J Jr, Hendrickson O Jr, Leonard R & Asmussen L (1984) Riparian forests as nutrient filters in agricultural watersheds. BioScience 34: 374–377

Luizao FJ (1995) Ecological Studies in Three Contrasting Forest Types in Central Amazonia. PhD Dissertation, University of Stirling, U.K.

Martinelli LA, Piccolo MC, Townsend AR, Vitousek PM, Cuevas E, McDowell W, Robertson GP, Santos OC & Treseder K. Nitrogen stable isotope composition of leaves and soil: tropical versus temperate forests. Biogeochemistry, this volume

Matson PA & Vitousek PM (1987) Cross-system comparisons of soil nitrogen transformations and nitrous oxide flux in tropical forest ecosystems. Global Biogeochemical Cycles 1: 163–170

Matson PA & Vitousek PM (1990) Ecosystem approach for the development of a global nitrous oxide budget. Bioscience 40(9): 667–672

Matson PA, Billow CR, Hall SJ & Zachariassen J (1996) Nitrogen trace gas responses to fertilization in sugar cane ecosystems. J. Geophys. Res. 101(D13): 18522–18545

Matson PA, Gower ST, Volkmann C, Billow C & Grier CC (1992) Soil nitrogen cycling and nitrous oxide fluxes in fertilized Rocky Mountain Douglas-fir forests. Biogeochemistry 18: 101–117

Matson PA, Vitousek PM, Ewel JJ, Mazzarino MJ & Robertson GP (1987) Nitrogen transformations following tropical forest felling and burning on a volcanic soil. Ecology 68: 491–502

Matthews E (1994) Nitrogenous fertilizers: global distribution of consumption and associated emissions of nitrous oxide and ammonia. Global Biogeochemical Cycles 8: 411–439

McClain ME, Richey JE & Pimentel TP (1994) Groundwater nitrogen dynamics at the terrestrial-lotic interface of a small catchment in the central Amazon Basin. Biogeochemistry 27: 113–127

McDowell WH, Bowden WB & Asbury CE (1992) Riparian nitrogen dynamics in two geomorphologically distinct tropical rain forest watersheds – subsurface solute patterns. Biogeochemistry 18: 53–75

McDowell WH, McSwiney CP & Bowden WB (1996) Effects of hurricane disturbance on groundwater chemistry and riparian function in a tropical rain forest. Biotropica 28: 577–584

McNulty SG & Aber JD (1993) Effects of chronic nitrogen additions on nitrogen cycling in a high-elevation spruce-fir stand. Can. J. Forest Res. 23: 1252–1263

Medina E & Cuevas E (1989) Patterns of nutrient accumulation and release in Amazonian forests of the upper Rio Negro Basin. In: Proctor J (Ed.) Mineral Nutrinets in Tropical Forest and Savanna Ecosystems (pp 217–240). Blackwell Scientific, Oxford

Melillo JM, Steudler PA, Aber JD & Bowden RD (1989) Atmospheric deposition and nutrient cycling. In: Andreae MD & Schimel DS (Eds) Exchange of Trace Gases between Terrestrial Ecosystems and the Atmosphere. Springer-Verlag

Mogollon JL & Querales E (1995) Interactions between acid solutions and Venezuelan tropical soils. Sci. Total Env. 164: 45–56

Nadelhoffer K, Downs MR, Fry B, Aber J, Magill A & Melillo J (1995) The fate of ^{15}N-labelled nitrate added to a northern hardwood forest in eastern Maine, U.S.A. Oecologia 103: 292–301

Nepstad DC, de Carvalho CR, Davidson EA, Jipp PH, Lefebvre PA, Negreiros GH, da Silva ED, Stone TA, Trumbore SE & Vieira S (1994) The deep-soil link between water and carbon cycles of Amazonian forests and pastures. Nature 372: 666–669

NRC (1986) Acid Deposition: Long-term Trends. National Academy of Sciences Press, Washington, DC, U.S.A.

Papen H, Hellmann B, Papke H & Rennenberg H (1993) Emissions of N-oxides from acid irrigated and limed soils of a coniferous forest in Bavaria. In: Oremland R (Ed.) Biogeochemistry of Global Change: Radiatively Active Trace Gases. Chapman and Hall, NY, U.S.A.

Pardo MT, Guadalix ME & Garcia-Gonzalez MT (1992) Effect of pH and background electrolyte on P sorption by variable charge soils. Geoderma 54: 275–284

Parton WJ, Stewart JWB & Cole CV (1988) Dynamics of C, N, P, and S in grassland soils: a model. Biogeochemistry 5: 109–131

Parton WJ, Mosier AR, Ojima DS, Valentine DW, Schimel DS, Weier K & Kulmala AE (1995) Generalized model for N_2 and N_2O production from nitrification and denitrification. Global Biogeochemical Cycles10(3): 401–412

Payne AI (1986) The Ecology of Tropical Lakes and Rivers. John Wiley and Sons, NY, U.S.A.

Peterjohn WT, Adams MB & Gilliam FS (1996) Symptoms of nitrogen saturation in two central Appalachian hardwood forests. Biogeochemsitry. 35(3): 507–529

Peterson BJ & Melillo JM (1985) The potential storage of carbon caused by eutrophication of the biosphere. Tellus 37B: 117–127

Pringle CM, Paaby-Hansen P, Vaux PD & Goldman CR (1986) *In situ* nutrient assays of periphyton growth in a lowland Costa Rican stream. Hydrobiologia 134: 207–213

Proctor J (Ed.) (1989) Mineral Nutrients in Tropical Forest and Savanna Ecosystems. Blackwell Scientific, Oxford

Reuss JO & Johnson DW (1986) Acid Deposition and the Acidification of Soils and Waters. Ecological Studies Publ. 59. Springer Verlag, New York, U.S.A.

Sanchez PA (1981) Soils of the humid tropics. In Department of Anthropology (Ed) Blowing in the Wind: Deforestation and Long-range Implications (pp 347–410). College of William and Mary, Williamsburg, VA, U.S.A.

Sanchez PA, Bandy DE, Villachica JH & Nicholaides JJ III (1982) Amazon Basin soils: Management for continuous crop production. Science 216: 821–827

Sanchez PA & Logan JT (1992) Myths and science about the chemistry and fertility of soils in the tropics. In: Lal R & Sanchez PA (Eds) Myths and Science of Soils in the Tropics, SSSA Special Publication 29. ASA and SSSA, Madison, WI, U.S.A.

Schimel DS (1995) Terrestrial ecosystems and the carbon cycle. Global Change Biology 1: 77–91

External inputs

Atmospheric deposition, geological weathering, and N-fixation are the principal potential sources of external N input to forest ecosystems. The N content of bulk precipitation for the HTF was estimated as the sum of measured NO_3-N and NH_4-N concentrations (McDowell et al. 1990) and an estimate of dissolved organic nitrogen (DON). Inputs of dissolved organic nitrogen (DON) have not been quantified for any HTF locations, but an estimate of DON was generated using DOC:DON ratios for precipitation at La Selva, Costa Rica (Eklund et al. 1997) and measured DOC inputs in HTF rain (McDowell et al. 1990). Bedrock N content has not been measured at any HTF locations; however, N concentrations have been measured to 6 m (Scatena, pers. comm.), a depth which greatly exceeds that of the HTF rooting zone (< 1 m). Edmisten (1970) report elevated rates of N-fixation for specific species of trees and epiphytes with high N-fixation potential. However, there are no published data for spatially-integrated average levels of N-fixation above- or belowground in HTF stands. Numerous authors have suggested that N-fixation may be an important source of "missing N"; we address the circumstantial evidence surrounding that hypothesis in the Discussion.

Outflows

Aqueous export and gaseous losses are the principal pathways for N loss from HTF ecosystems. We assume that concentrations of N entering the system in lateral subterranean flow are the same as those in lateral flow leaving the site. Although we assume that the concentration is constant, total water flux increases and hence there is a net loss of N from the plot proportional to runoff generation in the plot. Nitrogen flux from groundwater was calculated as the product of annual basin-wide runoff for the HTF (García-Martinó et al. 1996, McDowell & Asbury 1994) and total N concentrations in shallow subterranean groundwater of the HTF area in the Bisley watershed (McDowell et al. 1992). We use groundwater instead of streamwater N concentrations to estimate inorganic aqueous export in order to avoid the potentially confounding effects of riparian and in-stream N processing. Although riparian zones can alter hydrologic estimates of runoff generation due to increased evapotranspiration, they only occupy 7% of the watershed area (Scatena & Lugo 1995) and therefore do not have a significant impact on water balance in this system. Other potential complications, such as temporal variation in subterranean groundwater concentrations, do not significantly impact our estimate due to the lack of seasonal variability in groundwater in the HTF (McDowell et al. 1992). Annual particulate organic N export for a single HTF watershed averaged over a three-year period was used as an estimate of particulate N

Table 1. Fluxes of nitrogen (kg/ha/yr) in a hillslope tabonuco forest, Puerto Rico. Data reported as ranges of published values. Calculations are included in Appendix I.

	External Input	Outflows	Biomass Accumulation	Internal Flux
Precipitation[1]	4			
Dry Deposition	0–2[a]			
Geological Weathering	0[b]			
Nitrogen Fixation	?			
Total Inputs	4–6 + ?			
Groundwater[2,3]		5–11		
Particulate Organic N Losses[3]		0.3–0.7		
Gaseous N Losses[4,5]		1–4		
Total Outputs		6.3–15.7		
Aboveground Biomass Accumulation[6,7,8]			6	
Belowground Biomass Accumulation[6,7,8]			2	
Total Biomass Accumulation			8	
Whole Tree Mortality[6,9]				18
Coarse Woody Debris (Branchfall)[6,10]				6
Coarse Root Turnover[6,10]				1
Fine Litterfall[7,11]				101–103
Fine Root Turnover[6,12,13]				17
Net Throughfall[1,14,15]				3
Total N Uptake Requirement			154–156[c]	

[1]McDowell et al. 1990; [2]McDowell et al. 1992; [3]McDowell & Asbury 1994; [4]Erickson in prep.; [5]Data this study (Appendix II); [6]Scatena et al. 1993; [7]Weaver & Murphy 1990; [8]Crow 1980; [9]Lugo & Scatena 1996; [10]Vogt et al. 1996; [11]Lodge et al. 1991; [12]Silver & Vogt 1993; [13]Vogt et al. 1995; [14]McDowell 1998; [15]Scatena 1990.

[a]Estimates potential range based on dry deposition rates at other sites.

[b]Estimate based on basin geologic characteristics.

[c]Total N uptake requirement equals biomass accumulation + internal flux.

losses (McDowell & Asbury 1994). We reasoned that overall loss of particulate N from the basin was the best estimate of loss from HTF; we do not have the detailed plot-level studies of particulate N transport that would be needed to improve this estimate. Nitrous-oxide (N_2O) and nitric-oxide (NO) production rates were measured for three watersheds within the HTF zone (Erickson et al. in prep.; see Appendix II). We used acetylene inhibition of N_2 production to determine the ratio of N_2 to N_2O for the three watersheds

(Appendix II). We use this ratio as the basis for calculating total gaseous N loss from Erickson et al.'s (in prep.) measurements of N_2O production. NH_3 volatilization has not been measured in the LEF, but is generally considered to be small in forested ecosystems where NH_3 concentrations in ambient air are below the compensation point of 0.8 ppb (Langford & Fehsenfeld 1992) and soils are acidic.

Biomass accumulation

Principal reservoirs for N are aboveground biomass, root biomass, forest floor and soil organic matter (SOM). Nitrogen storage in aboveground biomass and roots in HTF stands is reported in Scatena et al. (1993). Nitrogen storage in forest floor and soils (0–60 cm) for the same locations are reported by Silver et al. (1994).

Following Scatena et al. (1993) we used Crow's (1980) estimates of aboveground woody biomass accumulation and Scatena et al.'s (1993) average aboveground woody biomass N concentrations to calculate net aboveground N sequestration for HTF. Biomass accumulation was estimated during an inter-hurricane period over the course of a 30-year period (Crow 1980). Recent data from Scatena et al. (1996) on hurricane damaged areas suggest this pre-hurricane estimate is still accurate. Because we know of no data on net accumulation of coarse root biomass in HTF we assumed an equal ratio of net accumulation to standing biomass for aboveground woody biomass and coarse root biomass to calculate a net belowground accumulation estimate. We then used the average coarse root N concentration (Scatena et al. 1993) to calculate an estimate of net N sequestration in coarse roots. We developed this N budget with the assumption that foliar, fine root, forest floor and SOM N reservoirs have reached steady-state conditions. Consequences of deviations from steady state are evaluated in the Discussion and Conclusion Sections.

Internal fluxes

We consider above- and belowground litterfall, litter decomposition, net SOM mineralization and net throughfall (total throughfall minus precipitation) to be the principal internal N fluxes in HTF stands. Ecosystem scientists in the LEF and elsewhere have adhered to an operational distinction between litterfall decomposition and net SOM mineralization. Although SOM mineralization is clearly part of the litterfall decomposition process, we maintain this operational distinction wherein N flux from leaf litter is accounted for as decomposition and N flux from SOM is estimated by determining potential net mineralization rates.

Aboveground litterfall includes leaves, fine wood < 1 cm diameter, and reproductive parts (Lodge et al. 1991; Weaver & Murphy 1990) in addition to coarse woody debris 1–6 cm diameter (Vogt et al. 1996). We used litterfall biomass reported in the references cited and multiplied these values by N concentrations given therein or in Scatena et al. (1993) to calculate N flux. Also included here is an estimate of N flux associated with whole tree mortality, which includes both above- and belowground components (Lugo & Scatena 1996; Scatena et al. 1993). We recognize that dead trees may remain standing for long periods and that net N immobilization rather than net N release characterizes early phases of dead wood decomposition. Nonetheless, the long-term fate of N sequestered in dead wood is solubilization (and subsequent uptake or loss) or transfer to the SOM N pool.

Belowground litterfall refers primarily to fine root turnover compiled from data reported in Scatena et al. (1993) and Silver & Vogt (1993). Because we do not have an estimate of fine root turnover under control or ambient conditions for HTF, we used a standard technique to generate this estimate. Fine root turnover was calculated following the method of McClaugherty et al. (1982) as the difference between annual maximum and minimum for dead root biomass (Silver & Vogt 1993) with N content of fine roots taken from Scatena et al. (1993; Table 1). Because we know of no data on belowground coarse root turnover, we estimate belowground coarse root turnover by assuming an equal ratio of aboveground coarse woody debris (branchfall) to aboveground woody biomass and coarse root biomass to calculate a turnover rate for coarse roots (Scatena et al. 1993; Vogt et al. 1996). We then used the average coarse root N concentration (Scatena et al. 1993) to calculate an estimate of N flux from this process. Live root N concentrations are used for estimating coarse and fine root turnover, therefore, retranslocation of N prior to senescence is included in these estimates. The belowground component of whole tree mortality was estimated by applying stem mortality rates (Lugo & Scatena 1996) to large coarse root biomass (Scatena et al. 1993).

Decomposition rates and associated net N releases from aboveground litterfall (Table 2) are derived from data presented by Weaver & Murphy (1990), Lodge et al. (1991), Vogt et al. (1996), Zimmerman et al. (1995), and Zou et al. (1995). Vogt et al. (1996) report rates of decomposition for coarse woody debris (1–6 cm diameter), and rates for large woody debris (Zimmerman et al. 1995) were used to estimate above- and belowground whole tree decay. Analogous data from fine root decay are provided by Silver & Vogt (1993) using *in situ* methods.

Data on N immobilization – mineralization rates during decomposition are lacking for most HTF plant tissue types. We used decay constants from mass loss and assumed that N loss was linearly proportional to mass loss. We recognize that this assumption is not valid particularly during the first stages of decomposition when N is immobilized. Zou et al. (1995) report rates of mass loss for decomposing leaf litter at 75.8% yr^{-1}, but their data on N loss from this litter translate into a N loss rate of approximately 50% during the first year of decomposition. We recognize that different plant tissues will immobilize and mineralize N at different rates, however litter types with the largest N contribution, such as fine litterfall and leaffall, also decompose the most rapidly. In addition, our estimates of litter turnover do not explicitly include N release from litter that fell during previous years. Because we are most interested here in the total annual N input from litterfall to the forest floor, which includes litterfall from previous years, we estimate N loss using mass loss decay constants without any adjustment for the initial immobilization phase.

Net mineralization of N from SOM was estimated using data from Steudler et al. (1991), Bowden et al. (1992), and Silver et al. (1994). Soil samples were taken from reference and hurricane plots from 0 to 2 cm depth and net mineralization rates were calculated as the net change in soil nitrate and ammonium concentrations over a seven-day period (Stuedler et al. 1991). Silver et al. (1994) report HTF SOM as 154 mg/ha from 0 to 60 cm depth. According to Bowden et al. (1992) soils from 0 to 10 cm in depth exhibited high rates of potential net N mineralization (average 14.3 μg N/ g soil/ day) while soils from 35 to 100 cm depth exhibited virtually no potential for net mineralization. We assume that $> 95\%$ of net N mineralization is occurring in the top 10 cm of soil where organic matter inputs are continual, and atmospheric inputs of water and oxygen are at a maximum. Silver et al. (1994) report percent organic matter and bulk densities for 3 soil depths (0–10 cm, 10–35 cm, and 35–60 cm); therefore, we can accurately determine the percentage of SOM present in the top 10 cm of soil for the Bisley forest and subsequently calculate net N mineralization as a function of available SOM (Table 2).

McDowell (1998) measured throughfall N flux (NH_4-N and NO_3-N) for El Verde; we estimate DON in throughfall by using the ratio of DOC:DON generated by Qualls & Haines (1991) and Qualls et al. (1991), and DOC concentrations from McDowell (1998). Net N transfer to the forest floor was estimated as the difference between canopy throughfall and precipitation. The quantity of annual canopy throughfall is relatively uniform through the HTF, approximately 59% of incoming precipitation (Scatena 1990); throughfall chemistry is only available for the El Verde site. Total annual watershed stemflow is reported in Scatena (1990). Some data suggest that N content in

Table 2. Decomposition and mineralization data for hillslope tabonuco forest, Puerto Rico. All input and decomposition values are reported in kg N/ha/yr and rounded to the nearest whole number. We applied the decay constant from mass loss (k) to N loss. Errors inherent in this assumption and our justification for making it are discussed in the Internal Fluxes portion of the Methods section.

	Total Input	Decomposition	k (yr^{-1})
Fine Litterfall[1,2]	101–103	78–80	0.78
Coarse Woody Debris (Branchfall)[3,4]	6	4	0.69
Fine Root Turnover[3,5,6]	17	7	0.40
Aboveground Tree Mortality[3,7]	13	1	0.10
Belowground Coarse Root Mortality[3,7]	5	1	0.10
Soil N Mineralization Estimate[8,9,10]		46–178	

[1]Lodge et al. 1991; [2]Weaver & Murphy 1990; [3]Scatena et al. 1993; [4]Vogt et al. 1996; [5]Silver & Vogt 1993; [6]Vogt et al. 1995;[7]Lugo & Scatena 1996; [8]Stuedler et al. 1991; [9]Silver et al. 1994; [10]Bowden et al. 1992.

stemflow is similar to throughfall (Edmisten 1970), but complete data on N content of stemflow are not available and thus are not included in our analysis.

Results and discussion

Results of our data analyses are summarized in Tables 1 and 2; we present a graphic illustration of the HTF N budget in Figure 2. Below we discuss these results and some of their implications.

External inputs

Bulk precipitation measurements for NH_4-N and NO_3-N (2 kg N/ha/yr) include wet deposition plus an unknown fraction of dry deposition. The inclusion of an estimate of DON in precipitation (2 kg/ha/yr) using DOC:DON ratios for rainfall in Costa Rica (8.2:1; Eklund et al. 1997) increased total precipitation N inputs two-fold. Dry deposition data were not available for any HTF stands. Estimates of throughfall suggest a minimal input of marine aerosols from dry deposition (McDowell 1998); however this does not necessarily translate into low levels of gaseous dry deposition of N. Lovett & Lindberg (1993) report N deposition data for 11 forests around the U.S. and one in Europe with dry deposition ranging from 39% to 56% of total N deposition. Based on the potential of long-range pollution transport to the LEF (McDowell et al.1990), precipitation inputs reported here could conceivably increase by 2 kg N/ha/yr with the inclusion of dry deposition.

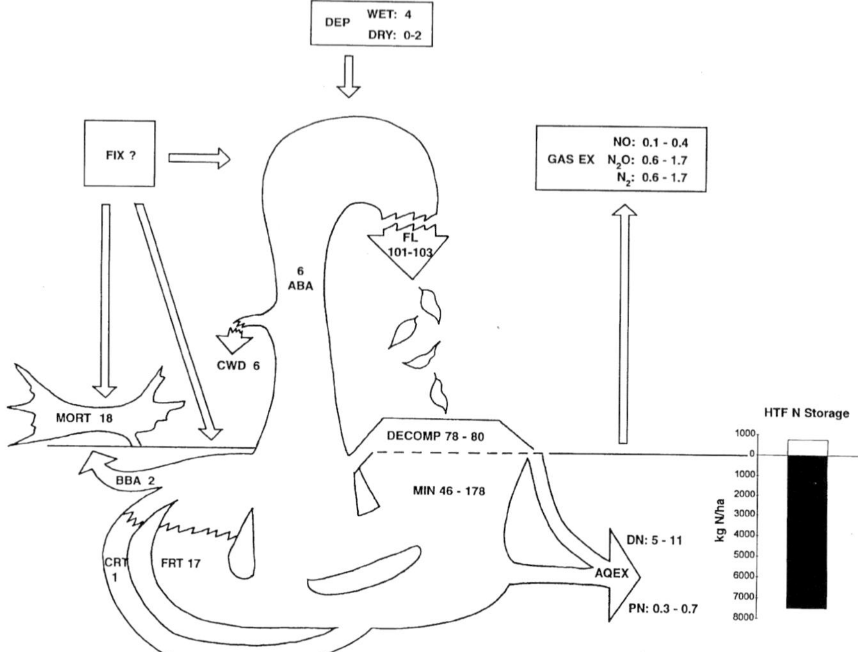

Figure 2. A model of the HTF N cycle. FIX – N-fixation, DEP – bulk deposition, GAS EX – gaseous, N export, FL – fine litterfall, ABA – aboveground biomass accumulation, CWD – coarse woody debris, MORT – above + belowground mortality, BBA – belowground biomass accumulation, CRT – coarse root turnover, FRT – fine root turnover, DECOMP – decomposition, MIN – net SOM mineralization, AQEX – aquesous export (groundwater + particulate organic N). All values are in kg N/ha/yr except for the accompanying bar graph, which illustrates HTF N pools (kg N/ha). Aboveground pools (empty bar) include forest floor (84 kg N/ha), understory vegetation (56 kg N/ha), woody tissue (506 kg N/ha) and foliage 108 kg N/ha); belowground pools (solid bar) include SOM (7250 kg N/ha), fine roots (35 kg N/ha) and coarse roots (203 kg N/ha).

Nitrogen inputs from geological weathering were not available for HTF stands. Some sedimentary rocks contain up to 0.4 percent N due to the incorporation of organic material into the rock matrix and their weathering can be an important N flux in some ecosystems (Dahlgren et al. 1994). However, volcaniclastic bedrock underlying the HTF probably contains little N, so weathering contribution is likely minimal. Measured N concentrations from 1–5 m depth in an HTF soil profile were always < 0.02 percent; below 5 m, N was undetectable (F.N. Scatena, pers. comm.).

Nitrogen fixation may be an important input of N into the HTF ecosystem. Edmisten (1970) reported the presence of several species of N-fixing bacteria and blue-green algae in HTF. To date no research has been published on the spatially-integrated rates of N-fixation in the canopy or forest floor for the HTF or other forest zones in Puerto Rico.

In most terrestrial ecosystems inputs from N-fixation are between 1 and 10 kg N/ha/yr (Cushon & Feller 1989; Perry 1994). However, levels of N-fixation are highly dependent on both plant and animal species composition. Several ecosystem types ranging from grasses to coniferous forests have exhibited high levels of potential N-fixation (50 to 150 kg/ha/yr) even without the presence of symbiotic N-fixing species (Bormann et al. 1993; Perry 1987; Stevenson 1986). Canopy lichens may contain N-fixing cyanobacteria that are capable in some cases of fixing large amounts of N (1.5 to 9 kg/ha/yr) in tropical forests (Forman 1975). Termites, which are very common in the LEF, may also be a host organism for N-fixing bacteria especially in tropical environments (Breznak & Brill 1973; Martius 1994; Perry 1994). Elevated rates of N-fixation are associated with primary succession and early pedogenesis. For the first 191 years of soil development on new volcanic substrates in Hawaii, N-fixation averaged 22 kg N/ha/yr (Vitousek et al. 1983). Hurricanes and landslides occur frequently in the LEF (Scatena & Lugo 1995); the latter disturbance is especially likely to trigger elevated rates of N-fixation (Zarin & Johnson 1995).

Outflows

We calculated a range for groundwater export (5–11 kg/ha/yr) which is similar to, though somewhat higher than, that of streamwater export found for a single HTF watershed by McDowell & Asbury (1994). Particulate organic N losses (0.3–0.7 kg N/ha/y)are also included to account for export due to surface erosion and runoff, but are considerably smaller than losses due to groundwater export.

Losses of nitrogen oxide (N_2O and NO; 0.6–1.7 kg N/ha/yr and 0.1–0.4 kg N/ha/yr, respectively) were measured for 3 late-successional HTF sites (Erickson in prep; see Appendix 2). While N_2 losses were not measured directly, we estimated losses of N_2 from these sites using the acetylene (C_2H_2) inhibition assay (Mosier & Klemedtsson, 1994; see Appendix 2). This assay relies upon the observation that, at high concentrations, C_2H_2 blocks the reduction of N_2O to N_2 during the process of denitrification. We followed the approach of Parsons et al. (1993) using intact soil cores. Randomly selected fresh soil cores were incubated for several hours with and without C_2H_2. We estimate the ratio of about 1:1 N_2 to N_2O by comparing the amount of N_2O produced under C_2H_2 to N_2O produced by control cores. Using this ratio, and assuming that all N_2O flux observed by Erickson et al. (in prep.) was due to denitrification, we calculate total gaseous N flux for HTF as 1–4 kg N/ha/yr. Other studies in HTF (Bowden et al.1992; Keller et al. 1986; and Stuedler et al. 1991) report N_2O fluxes similar to those of Erickson et al. (in prep); however, they do not include any losses of N as N_2 directly. Furthermore, Erickson et al. (in prep) include improved spatial replication and account for seasonal variation with their sampling strategy. The ranges we report in this budget are for N losses from NO, N_2O and N_2 calculated from the data of Erickson et al. (in prep.) and do not include data reported in previous studies. We note that NO emissions from the forest floor represent a maximum NO loss from the system. Kaplan et al. (1988) and Bakwin et al. (1990) observed that in Amazonian forests some or all of the NO produced at the soil surface could be recycled to the forest vegetation after chemical reaction to NO_2 and deposition on forest surfaces.

Biomass accumulation

Based on estimates of net accumulation of woody biomass, we know that the above- and belowground biomass pools are not in steady-state but rather there is a net sequestration of approximately 8 kg N/ha/y into the biomass pool annually. This calculation is based on long-term average biomass accumulation rates and a one-time measurement of plant tissue N concentrations. As such, we cannot specifically determine the actual N sequestration rate for any given year. The temporal dynamics of N accumulation in biomass are complex because, following catastrophic disturbance, N in aboveground biomass approaches pre-disturbance levels faster than the biomass itself (Scatena et al. 1996; Zarin & Johnson 1995). During succession, C:N ratios increase as the ratio of woody to nonwoody tissue increases. This proportional dilution of the biomass N pool implies that the curve of biomass N accumulation reaches its asymptote earlier than the asymptote for biomass accumulation itself is reached. Consequently, our method of calculating N accumulation in biomass sets an upper bound.

Internal fluxes

Annual above- and belowground litterfall turnover rates are quite high relative to other N fluxes (Table 1). These intrasystem processes supply the evergreen forest with much of the N necessary for plant growth and maintenance. Fine litter inputs (101–103 kg N/ha/yr) are the largest aboveground input of N to the forest floor. Compared to the external inputs and outflows, other forms of aboveground litter (branchfall and aboveground whole tree mortality) also provide a considerable amount of N to the forest floor annually (6 and 13 kg N/ha/yr, respectively). Some overlap may occur between these two measures because aboveground whole tree mortality may include some portion of branchfall. Belowground fine root turnover (17 kg N/ha/yr), coarse root turnover (1 kg/ha/yr) and belowground whole tree mortality (5 kg N/ha/yr) cycle considerably less N than their aboveground counterparts. However, these belowground sources are still significant relative to other N fluxes.

Decomposition rates for aboveground fine litter (78–80 kg N/ha/yr) are quite high relative to decay rates of fine roots and larger debris associated with whole tree mortality (Table 2; Silver & Vogt 1993; Vogt et al. 1996; Weaver & Murphy 1990; Zimmerman et al. 1995; Zou et al. 1995). The rapid decay rates reported here and the relatively large amount of net SOM-N mineralization suggest that cycling of N, especially from fine litter, occurs quite rapidly within this tropical forest system (Table 2).

Conclusions

Measured external inputs and outputs to the hillslope *tabonuco* forest are not balanced: inputs from precipitation (4 kg N/ha/yr) and geological weathering (~0 kg N/ha/yr) are considerably less than outputs from the system through groundwater (5–11 kg N/ha/yr), particulate organic N export (0.3–0.7 kg/ha/yr) and gaseous N loss (1–4 kg N/ha/yr). The long-term accretion in living biomass both aboveground and belowground sequesters up to 8 kg N/ha/yr in the hillslope forest, and this N must be supplied from additional unmeasured inputs or a change in N storage in SOM. Therefore, even with the inclusion of additional N from dry deposition (up to 2 kg/ha/yr), we have a considerable budgetary imbalance (8.3–19.7 kg N/ha/yr). It is unlikely that such a large imbalance is simply due to error associated with budget generation. Our inclusion of gaseous N loss and belowground biomass accumulation lead to a larger imbalance than the catchment-scale N deficit reported by McDowell and Asbury (1994).

There are several possible sources of this "missing N" (Bormann et al. 1977; Likens et al. 1977). First, while we assume that the SOM pool is in steady state, small and unmeasurable deviations from steady state within this large pool (7250 kg N/ha) could lead to large inputs of available N. Nitrogen could be surreptitiously supplied to the system either through an increase in soil mineralization or through a change in soil C:N ratios. A 1% increase in the soil C:N ratio over a ten-year period could provide over 7 kg/ha/yr of additional N to the system, which would account for much of the N budget imbalance. Johnson et al. (1995) have reported a >1 percent per year net loss of carbon from the solum in an undisturbed late-successional northern hardwoods watershed, an indication that the prevalent assumption of a steady-state SOM pool may be unwarranted. Based on the size of the HTF SOM-N pool, an annual drawdown of ~0.1 percent of the SOM-N could account for most of our estimated N imbalance.

The size of our estimated imbalance between known inputs and outputs plus system accretion requirements is large relative to the actual input and output measurements. Since N-fixation may be actively occurring within the HTF (Edmisten 1970), unmeasured N-fixation is likely to account for some portion of our N imbalance. The lack of analyses of above- and belowground N-fixation and limitations inherent in published measurements of internal solum N fluxes in HTFs constrain our ability to do more than speculate about the potential contributions of N-fixation and net SOM-N mineralization to the N cycle of this forest type.

Finally, we cannot exclude the impacts of experimental error due to inter-annual variability in individual fluxes. We have compiled data collected over a number of years at several HTF sites, thereby enhancing the potential for error. Principal limitations of the flux measurements are largely related to the spatial heterogeneity and temporal variability inherent in these processes (Robertson et al. 1988).

Spatial heterogeneity

Microsite variability within HTF stands may lead to significant differences in estimations of gaseous exchange and net mineralization. Robertson et al. (1988) determined that, within a Michigan forest, the spatial heterogeneity of N flux processes (denitrification, nitrification, and mineralization) was complex and dependent upon surface topography. For tropical dry forests various aspects of the N cycle have also been found to be affected by topography and microsite variability (Roy & Singh 1994, 1995; Raghubanshi 1992). Topographic variability led to heterogeneous aboveground litter accumulation and fine root production, causing alterations in net mineralization rates and net primary productivity (Roy & Singh 1994; 1995). Nitrogen mineralization and nitrification were found to decrease along a topographic gradient in Michigan forests (Raghubanshi 1992), and Silver et al. (1994) report that other soil characteristics vary along a topographic gradient at the Bisley watershed. We expect that the highly complex and varied surface topography of HTF ecosystems also produces microsite variability in N flux rates, as it does in the riparian zones of the Luquillo Forest (McDowell et al. 1992; Bowden et al. 1992).

Temporal variability

The LEF is heavily influenced by hurricanes with 83 percent of the landscape being affected each century (Lugo & Scatena 1996). Although hurricanes are the most extensive disturbance in the LEF, landslides and tree fall gaps affect the forest as well, with landslides having a recurrence interval of 3,000 to

10,000 years, affecting 3 percent of the landscape each century (Walker et al. 1996). Increased N inputs resulting from large scale disturbance may in fact be offset by the increased N requirements of forest regeneration in the LEF. Aboveground N pools in biomass following Hurricane Hugo in 1989 decreased initially but returned to pre-hurricane levels within 48 months due to the rapid growth rates of regenerating vegetation (7–10 times the growth rate of mature forest) and the high N content of new wood and herbaceous plants (Scatena et al. 1996). Following harvest and hurricane disturbances, Silver et al. (1996) found that belowground stocks of NO_3 and NH_4 were altered, but effects were short-term and the system returned to pre-disturbance levels within one year. McDowell et al. (1996) report that inorganic N concentration in groundwater also returned to pre-hurricane levels in HTF within 1–2 yrs. after the disturbance. Temporal variability in biomass C:N ratios associated with successional development also contributes to changes in N cycling rates as detrital inputs to the SOM pool increasingly reflect the high C:N ratios characteristic of woody debris. Although disturbances may produce major alterations in forest composition and structure, the overall nitrogen economy appears both resistant and resilient to disturbance.

Directions for future research

Development of this N budget has led us to identify several key areas for future investigations. (1) Nitrogen fixation has not been studied on a large scale within the natural forest system in either canopy or soil environments and may be a significant source of N to this system. What is the magnitude, distribution, and spatial and temporal variability in N-fixation in the tabonuco forest? (2) Nitrogen within HTF stands appears to be cycled through decomposition and biotic uptake with internal fluxes considerably larger than losses. Are the SOM and forest floor pools really in steady-state? (3) Temporal variability may be high for mineralization and gaseous exchange based on the disturbance regime of the watershed; however, nitrogen reserves above- and belowground appear to stabilize quickly. How are the impacts of large (hurricanes and landslides) and small (treefall gaps) disturbances on these N processes distributed within affected HTF stands? (4) Spatial and temporal variability may lead to misinterpretations especially with respect to mineralization and gas exchange rates. What spatial and temporal scales of analysis are required to accurately represent these processes in stand and catchment-level nutrient budget analyses?

Acknowledgements

We thank H. Erickson for providing additional data for budget generation, F.N. Scatena, A.E. Lugo, and W.L. Silver for insightful comments on the manuscript, and L. Isaacson for graphical design of figures. This research was performed under grant DEB-9411973 from the National Science Foundation to the Terrestrial Ecology Division, University of Puerto Rico, and the International Institute of Tropical Forestry as part of the Long-term Ecological Research Program in the Luquillo Experimental Forest; along with the support of the Forest Service (U.S. Department of Agriculture) and in cooperation with the University of Puerto Rico. Additional financial support was provided from a grant to D.J. Zarin from the Andrew W. Mellon Foundation. We acknowledge support from the NASA Terrestrial Ecology Program (NAGW-3772) and the NASA-IRA program.

Appendix I

Values reported here and in Table 1 have been rounded to reflect our best estimate of precision.

1. **Precipitation = 4 kg N/ha/yr**
   NO_3-N + NH_4-N in bulk precipitation for El Verde = 1.95 kg N/ha/yr (McDowell et al. 1990)
   DON:DOC Ratio in precipitation at La Selva, Costa Rica = 1:8.2 (Eklund et al. 1997)
   DOC in precipitation at El Verde = 19.1 kg/ha/yr (McDowell et al. 1990)
   DON in precipitation at El Verde using ratio = 2.3 kg N/ha/yr
   N Flux = 1.95 + 2.3 = 4.25 kg N/ha/yr

2. **Groundwater = 5–11 kg N/ha/yr**
   Basin-wide runoff (Toronja) = 1750 mm/yr (McDowell & Asbury 1994)
   Basin-wide runoff (Bisley) = 1776 mm/yr (Scatena pers. comm.)
   Average basin-wide runoff (Tabonuco Forest Zone) = 1830 mm/yr (García-Martinó et al. 1996)
   Groundwater TDN concentration for Bisley slope = 0.3–0.6 g N/m^3 McDowell et al. 1992)
   Minimum N Flux from GW = 1750 mm/yr * 0.3 g N/m^3 = 5.3 kg N/ha/yr
   Maximum N Flux from GW = 1830 mm/yr * 0.6 g N/m^3 = 11.0 kg N/ha/yr
   N Flux - 5.3–11.0 kg N/ha/yr

3. **Particulate Organic N Export = 0.3–0.7 kg N/ha/yr**
   Range of basin-wide stream export for Toronja watershed for 3 years (McDowell & Asbury 1994)

4. **Denitrification = 1–4 kg N/ha/yr**
   N_2O and NO data from Erickson et al. in prep. for HTF
   El Verde N_2O = 0.75 ng N/cm^2/h; NO = 0.11 ng N/cm^2/h

Mameyes N_2O = 0.64 ng N/cm^2/h; NO = 0.14 ng N/cm^2/h
SSabana N_2O = 1.93 ng N/cm^2/h; NO = 0.40 ng N/cm^2/h
N_2 Ratio for HTF = 1:1 (this paper; Appendix II)
Minimum N Flux from N_2O and N_2 = 0.6 kg N/ha/yr * 2 = 1.2 kg N/ha/yr
Maximum N Flux from N_2O and N_2 = 1.7 kg N/ha/yr * 2 = 3.4 kg N/ha/yr
Minimum N Flux (N_2O, N_2 and NO) = 1.2 kg N/ha/yr + 0.10 kg N/ha/yr = 1.3 kg N/ha/yr
Maximum N Flux (N_2O, N_2 and NO) = 3.4 kg N/ha/yr + 0.35 kg N/ha/yr = 3.75 kg N/ha/yr
N Flux = 1.3–3.75 kg N/ha/yr

5. Aboveground Biomass Accumulation = 6 kg N/ha/yr

Net woody production Tabonuco Forest = 2268 kg/ha/yr (Crow 1980; Weaver & Murphy 1990)
Total N Bisley slope (aboveground vegetation) = 2.82 mg N/g biomass (Scatena et al. 1993)
N flux = 2268 kg/ha/yr * 0.00282 g/g = 6.4 kg N/ha/yr

6. Belowground Biomass (Coarse Roots) Accumulation = 2 kg N/ha/yr

Above biomass/net woody production = Below biomass/below woody production
Below biomass (coarse roots) = 65681 kg/ha (Scatena et al. 1993)
Above biomass (leaves, branches, bark, bole) = 224841 kg/ha (Scatena et al. 1993)
Net woody production = 2268 kg/ha/yr (Crow 1980; Weaver & Murphy 1990)
Below production = (73558 * 2268) / 224841 = 742 kg/ha/yr
Total N for Bisley slope (coarse roots) = 0.0031 g N/ g biomass (Scatena et al. 1993)
N flux = 0.0031 g/g * 742 kg/ha/yr = 2.4 N/ha/yr

7. Aboveground Fine Litterfall Turnover = 101–103 kg N/ha/yr

Bisley watershed mean annual nutrient input from total fine litterfall (Lodge et al. 1991)
Litterfall biomass = 9697 kg/ha/yr (Lodge et al. 1991)
Leaf biomass = 1.29 g/m^2/day; Leaf N content = 12.6 mg/g
Fine wood biomass = 0.55 g/m^2/day; Fine wood N content = 8.6 mg/g
Misc. fine litter biomass = 0.54 g/m^2/day; Misc. fine litter N content = 13.4 mg/g
N Flux = (1.29 g/m^2/day * 0.0126g/g) + (0.55 g/m^2/day * 0.0085g/g) + (0.54 g/m^2/day * 0.0134g/g) = 102.9 kg N/ha/yr
Litterfall biomass = 8748 kg/ha/yr (Weaver & Murphy 1990)
Avg N content = 0.0115 g N/g (Lodge et al. 1991)
N Flux = 8748 kg/ha/yr * 0.0115 g N/g = 100.6 kg N/ha/yr
N Flux = 100.6–102.9 kg N/ha/yr

8. Fine Root Turnover = 17 kg N/ha/yr

Maximum Dead Root Biomass = 175 g/m^2 (Silver & Vogt 1993)
Minimum Dead Root Biomass = 75 g/m^2 (Silver & Vogt 1993)
Annual Difference = 1000 kg/ha/yr
Total N Fine Roots = 0.017 g/g (Scatena et al. 1993)
N Flux = 1000 kg/ha/yr * 0.017 g N/g biomass = 17.0 kg N/ha/yr

Keller M & Reiners WA (1994) Soil-atmosphere exchange of nitrous oxide, nitric oxide, and methane under secondary succession of pasture to forest in the Atlantic Lowlands of Costa Rica. Global Biogeochemistry 8: 399–409

Langford AO & Fehsenfeld FC (1992) Natural vegetation as a source or sink for atmospheric ammonia: A case study. Science 255: 581–583

Likens GE, Bormann FH, Pierce RS, Eaton JS & Johnson NM (1977) Biogeochemistry of a Forest Ecosystem. Spingler-Verlag, New York, New York, U.S.A.

Lodge DJ, Scatena FN, Asbury CE & Sanchez MJ (1991) Fine litterfall and related nutrient inputs resulting from Hurricane Hugo in subtropical wet and lower montane rain forests of Puerto Rico. Biotropica 23: 336–342

Lovett GM & Lindberg SE (1993) Atmospheric deposition and canopy interactions of nitrogen in forests. Can. J. Forest Res. 23: 1603–1616

Lugo AE & Scatena FN (1996) Background and catastrophic tree mortality in tropical moist, wet and rain forests. Biotropica 28: 585–599

Martius C (1994) Diversity and ecology of termites in Amazonian forests. Pedobiologia 38: 407–428

McClaugherty CA, Aber JD & Melillo JM (1982) The role of fine roots in the organic matter and nitrogen budgets of two forested ecosystems. Ecology 63: 1481–1490

McDowell WH, Bowden WB & Asbury CE (1992) Riparian nitrogen dynamics in two geomorpholoically distinct tropical rain forest watersheds: subsurface solute patterns. Biogeochemistry 18: 53–75

McDowell WH & Asbury CE (1994) Export of carbon, nitrogen, and major ions from three tropical montane watersheds. Limnology and Oceanography 39: 111–125

McDowell WH, Gines-Sanchez C, Asbury CE & Ramos-Perez CR (1990) Influence of sea salt aerosols and long range transport on precipitation chemistry at El Verde, Puerto Rico. Atmospheric Environment 24A: 2813–2821

McDowell WH (1998) Internal nutrient fluxes in a tropical rain forest. J. Tropical Ecology 14: 521–536

McDowell, WH, McSwiney, CP & Bowden WB (1996) Effects of hurricane disturbance on groundwater chemistry and riparian function in a tropical rain forest. Biotropica 28: 577–584

Mitchell MJ, Foster NW, Shepard JP & Morrison IK (1992) Nutrient cycling in Huntington Forest and Turkey Lakes deciduous stands: Nitrogen and sulfur. Can. J. Forest Res. 22: 457–464

Mitchell MJ, Raynal DJ & Driscoll CT (1996) Biogeochemistry of a forested watershed in the central Adirondack Mountains: Temporal changes and mass balance. Water Air and Soil Pollution 88: 355–369

Monk CD & Day FP Jr (1988) Biomass, primary production, and selected nutrient budgets for an undisturbed watershed. In Swank WT & Crossley DA Jr (Eds) Forest Hydrology and Ecology at Coweeta (pp 151–159). Springer-Verlag, New York, U.S.A.

Mosier AR & Klemedtsson L (1994) Measuring denitrification in the field. In: Weaver RW, Angle S, Bottomley P, Bezdicek D, Smith S, Tabatabai A, and Wollum A (Eds) Methods of Soil Analysis: Part 2, Microbiological and Biochemical Properties (pp 1047–1065). Soil Science of America, Madison, Wisconsin, U.S.A.

Parsons WFJ, Mitre ME, Keller M & Reiners WA (1993) Nitrate limitation of N2O production and denitrification from tropical pasture soils. Biogeochemistry 22: 179–193

Perry DA, Chogquette C & Schroeder P (1987) Nitrogen dynamics in conifer-dominated forests with and without hardwoods. Can. J. Forest Res. 17: 1434–1441

Perry DA (1994) Forest Ecosystems. Johns Hopkins Univ. Press, Baltimore, MD, U.S.A.

Qualls RG & Haines BL (1991) Geochemistry of dissolved organic nutrients in water percolating through a forest ecosystem. Soil Science Society of America J. 55: 1112–1123

Qualls RG, Haines BL & Swank WT (1991) Fluxes of dissolved organic nutrients and humic substances in deciduous forest. Ecology 72: 254–266

Raghubanshi AS (1992) Effect of topography on selected soil properties and nitrogen mineralization in a dry tropical forest. Soil Biology and Biochemistry 24: 145–150

Robertson GP, Huston MA, Evans FC & Tiedje JM (1988) Spatial variability in a successional plant community: Patterns of nitrogen availability. Ecology 69: 1517–1524

Roy S & Singh JS (1995) Seasonal and spatial dynamics of plant-available N and P pools and N-mineralization in relation to fine roots in a dry tropical forest habitat. Soil Biology and Biochemistry 27: 33–40

Roy S & Singh JS (1994) Consequences of habitat heterogeneity for availability of nutrients in a dry tropical forest. J. Ecology 82: 503–509

Scatena FN & Lugo AE (1995) Geomorphology, disturbance, and the soil and vegetation of two subtropical wet steepland watersheds of Puerto Rico. Geomorphology 13: 199–213

Scatena FN, Silver W, Siccama T, Johnson A & Sanchez MJ (1993) Biomass and nutrient content of the Bisley experimental watersheds, Luquillo Experimental Forest, Puerto Rico, before and after Hurricane Hugo, 1989. Biotropica 25: 15–27

Scatena FN (1990) Watershed scale rainfall interception on two forested watersheds in the Luquillo Mountains of Puerto Rico. J. Hydrology 113: 89–102

Scatena FN, Moya S, Estrada C & Chinea JD (1996) The first five years in the reorganization of aboveground biomass and nutrient use following Hurricane Hugo in the Bisley Experimental Watersheds, Luquillo Experimental Forest, Puerto Rico. Biotropica 28: 424–440

Silver WL & Vogt KA (1993) Fine root dynamics following single and multiple disturbances in a subtropical wet forest ecosystem. J. Ecology 81: 729–738

Silver WL, Scatena FN, Johnson AH, Siccama TG & Sanchez MJ (1994) Nutrient availability in a montane wet tropical forest: Spatial patterns and methodological considerations. Plant and Soil 164: 129–145

Silver WL, Scatena FN, Johnson AH, Siccama TG & Watt F (1996) At what scale does disturbance affect belowground nutrient pools? Biotropica 28: 441–457

Sollins P, Grier CC, McCorison FM, Cromack K Jr & Fogel R (1980) The internal element cycles of an old-growth Douglas-Fir ecosystem in western Oregon. Ecological Monographs 50: 261–285

Steudler PA, Melillo JM, Bowden RD, Castro MS & Lugo AE (1991) The effects of natural and human disturbances on soil nitrogen dynamics and trace gas fluxes in a Puerto Rican wet forest. Biotropica 23: 356–363

Stevenson FJ (1986) Cycles of Soil. John Wiley & Sons. New York, U.S.A.

Triska FJ, Sedell JR, Cromack K Jr, Gregory SV & McCorison FM (1984) Nitrogen budget of a small coniferous forest stream. Ecological Monographs 54: 119–140

Veldkamp E & Keller M (1997) Nitrogen oxide emissions form a banana plantation in the humid tropics. J. Geophys. Res. 102: 15889–15898

Vitousek PM, Van Cleve K, Balakrishnan N, Mueller Dombois D (1983) Soil development and nitrogen turnover in montane rainforest soils in Hawaii. Biotropica 15: 268–274

Vogt KA, Vogt DJ, Asbjorsen H & Dahlgren RA (1995) Roots, nutrients and their relationship to spatial patterns. Plant and Soil 168–169: 113–123

Vogt KA, Vogt DJ, Boon P, Covich A, Scatena FN, Asbjorsen H, O'Hara JL, Siccama TG, Bloomfield J & Ranciato JF (1996) Litter dynamics along stream, riparian, and upslope areas following Hurricane Hugo, Luquillo Experimental Forest, Puerto Rico. Biotropica 28: 458–470

Walker LR, Brokaw NVL, Lodge DJ & Waide RB (1991) Ecosystem, plant, and animal responses to hurricanes in the Carribean. Biotropica 23: 313–521

Walker LR, Zarin DJ, Fetcher N, Myster RW & Johnson AH (1996) Ecosystem development and plant succession on landslides in the Caribbean. Biotropica 28: 566–576

Weaver PL & Murphy PG (1990) Forest Structure and Productivity in Puerto Rico's Luquillo Mountains. Biotropica 22: 69–82

Zarin DJ & Johnson AH (1995) Nutrient accumulation during primary succession in a montane tropical forest, Puerto Rico. Soil Science Society of America J. 59: 1444–1452

Zimmerman JK, Pulliam WM, Lodge DJ, Quinones-Orfila V, Fetcher N, S. Guzman-Grajales, Parrota JA, Asbury CE, Walker LR & Waide RB (1995) Nitrogen immobilization by decomposing woody debris and the recovery of tropical wet forest from hurricane damage. Oikos 72: 314–322

Zou X, Zucca CP, Waide RB & McDowell WH (1995) Long-term influence of deforestation on tree species composition and litter dynamics of a tropical rain forest in Puerto Rico. Forest Ecology and Management 78: 147–157

Biogeochemistry **46:** 109–148, 1999.

The impact of accelerating land-use change on the N-Cycle of tropical aquatic ecosystems: Current conditions and projected changes

J.A. DOWNING[1], M. McCLAIN[2], R. TWILLEY[3], J.M. MELACK[4], J. ELSER[5], N.N. RABALAIS[6], W.M. LEWIS, JR.[7], R.E. TURNER[8], J. CORREDOR[9], D. SOTO[10], A. YANEZ-ARANCIBIA[11], J.A. KOPASKA[1] & R.W. HOWARTH[12]

[1]*Department of Animal Ecology, Iowa State University, Ames, IA, U.S.A.;* [2]*RSMAS-MGG, University of Miami, Miami, FL, U.S.A.;* [3]*Department of Biology, University of Southwestern Louisiana, Lafayette, LA, U.S.A.;* [4]*Department of Ecology, Evolution and Marine Biology, University of California, Santa Barbara, CA, U.S.A.;* [5]*Department of Zoology, Arizona State University, Tempe, AZ, U.S.A.;* [6]*Louisiana Universities Marine Consortium, Chauvin, LA, U.S.A.;* [7]*Center for Limnology, Cooperative Institute for Research in Environmental Sciences, University of Colorado, Boulder, CO, U.S.A.;* [8]*Department of Oceanography and Coastal Sciences, Louisiana State University, Baton Rouge, LA, U.S.A.;* [9]*Department of Marine Science, University of Puerto Rico, Mayaguez, PR;* [10]*Facultad de Pesquierías y Oceanografía, Instituto de Ciencias Naturales y Exactas, Universidad Austral de Chile, Puerto Montt, Chile;* [11]*Institute of Ecology A.C., Department of Coastal Resources, Xalapa, Veracruz, Mexico;* [12]*Section of Ecology and Systematics, Cornell University, Ithaca, NY, U.S.A.*

Received 10 December 1998

Key words: estuaries, lakes, marine, nitrogen, phosphorus, rivers, streams, temperate, tropics

Abstract. Published data and analyses from temperate and tropical aquatic systems are used to summarize knowledge about the potential impact of land-use alteration on the nitrogen biogeochemistry of tropical aquatic ecosystems, identify important patterns and recommend key needs for research. The tropical N-cycle is traced from pre-disturbance conditions through the phases of disturbance, highlighting major differences between tropical and temperate systems that might influence development strategies in the tropics. Analyses suggest that tropical freshwaters are more frequently N-limited than temperate zones, while tropical marine systems may show more frequent P limitation. These analyses indicate that disturbances to pristine tropical lands will lead to greatly increased primary production in freshwaters and large changes in tropical freshwater communities. Increased freshwater nutrient flux will also lead to an expansion of the high production, N- and light-limited zones around river deltas, a switch from P- to N-limitation in calcareous marine systems, with large changes in the community composition of fragile mangrove and reef systems. Key information gaps are highlighted, including data on mechanisms of nutrient transport and atmospheric deposition in the tropics, nutrient and material retention capacities of tropical impoundments, and N/P coupling and stoichiometric impacts of nutrient supplies on tropical aquatic communities. The current base of biogeochemical data suggests that alterations in the cultural fields is greater in tilled fields than in pastures (Neill et al. 1995) and is greater in sandy soils than in clayey soils. When expressed as a percentage of nitrogen added in fertilizer, the nitrogen leaching from agricultural soils in surface and groundwaters varies from 28% to 85% (Hölscher et al. 1997; Kühne 1993; Poss & Saragoni 1992). Even when agricultural soils are not fertilized, they leach more nitrogen than do forests, perhaps due to lowered evapotranspiration and greater water runoff (Uhl & Jordan 1984; Seyfried & Rao 1991; Sur et al. 1992; Vitousek & Melillo 1979). When runoff water from tropical agricultural systems runs through wetlands before reaching streams, a large proportion of the leached nitrogen can be denitrified (Arheimer & Wittgren 1994; Peterjohn & Correll 1984; Pinay & Decamps 1988; Vought et al. 1994). Very high rates of denitrification may occur in the hyporheic zones of tropical streams (Sjodin et al. 1997). In the absence of substantial wetland buffer strips or hyporheic denitrification, however, levels of nitrate in streams may rise to levels well above those required by freshwater biota, so aquatic communities may be greatly altered by intensive agriculture (Pedrozo & Bonetto 1989). Physical practices related to intensive land cultivation also increase erosion and, therefore, further augment exports of particulate N.

In addition to nitrogen leaching from agricultural fields, agriculture can contribute to nitrogen mobilization in the landscape through increased volatilization to the atmosphere and subsequent deposition, particularly of NH_x. Howarth (1998) has demonstrated that at the scale of large regions, NH_x deposition is strongly correlated with riverine export of nitrogen, and that NH_x deposition appears to be a strong surrogate measure of the leakiness of "surplus" nitrogen from agricultural systems.

Phase 4 – *Water Management Systems* – Reservoir development often accompanies the initiation of large-scale intensive agriculture (Matsumura-Tundisi et al. 1981). It is also associated with increasing urban populations, industrialization, and hydroelectric power needs in the tropics (Tundisi 1981, 1994). The impact of reservoir development should generally lead to a decrease of nutrient exports from catchments (Bonetto et al. 1994; Pedrozo et al. 1992; Tundisi & Matsumura-Tundisi 1981; Vorosmarty et al. 1997). Reservoirs increase water residence times, thus allowing greater biotic uptake by aquatic vegetation, and the establishment of anoxic bottom water conditions that promote denitrification, acting to reduce exports of N to downstream receiving waters. Nutrient retention may be extremely dynamic, however, since some tropical lakes apparently retain nutrients during periods of low water but release them during the rainy season (Furch & Junk 1993). Experience in temperate ecosystems (Fleisher & Stibe 1991) suggests that N retention in lakes and natural wetlands varies between about 10% and 90% (including biotic uptake, sedimentation, and denitrification), depending upon

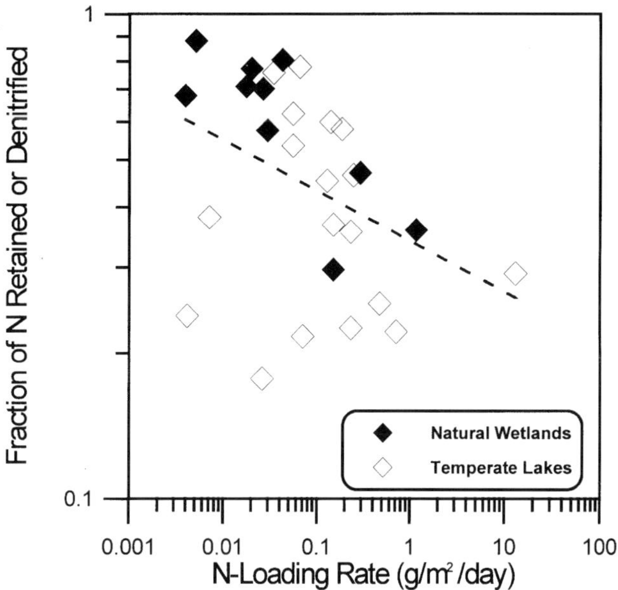

Figure 6. Relationship between nitrogen "retention" (including denitrification losses) and nitrogen loading rate in temperate zone lakes and ponds (Fleisher & Stibe 1991). The dashed line indicates the least squares regression relationship between the logarithms of the two variables ($n = 28$, $r = 0.39$, $p < 0.05$).

the hydraulic and biotic characteristics of the water body (Howarth 1998). Wetlands have very high rates of N retention (Figure 6), but N retention decreases as N-loading rises above about 0.1 g $m^{-2}day^{-1}$ (Figure 6). Although the data plotted in Fig. 6 may be somewhat optimistic compared to riverine N retention (Day & Kemp 1985), temperate data suggest that impoundments and wetlands can retain large amounts of N, but that their N removal efficiency will decrease as N loading rises. The impact of tropical impoundments on N loads may be less than impoundments in temperate systems, however, if N fixation rates in tropical waters are very high, or if high turbidity hampers biotic nutrient uptake (e.g., Arcifa et al. 1981b).

Phase 5 – *Major Urban and Industrial Development* – Within major urban and industrial areas, natural cycles of N are nearly completely overwhelmed by anthropogenic sources. High inputs of human waste and effluent from industrial activities enrich rivers with organic and inorganic nitrogen. Where light or turbulence do not renew oxygen, river waters may become hypoxic and the river biota may shift toward species tolerant of low oxygen conditions. Even the nutrient retention capacity of impoundments may be hampered because of very high rates of cultural eutrophication that accompanies urbanization in the tropics (Tundisi 1987; Arcifa et al. 1981a).

Without remedial measures, river exports of N will increase in proportion to anthropogenic inputs, unless denitrification is very rapid.

In large tropical river basins, several of these phases are likely to be active simultaneously in the landscape. Consequently, the downstream effects of each will be integrated into a single export to coastal waters. The magnitude of this integrated export will depend partially upon the proportion of the basin subject to each phase of development.

Impacts of tropical development on the relative availability of N and P

It is now well known that N, P, or other nutrients (Downing 1997) can limit primary production in freshwater and marine systems, and that the limiting nutrient is determined by the relative availability of ambient N, P and other essential nutrients. Therefore, understanding the expected changes in ecosystem function due to alteration of N cycling must include a consideration of accompanying alterations in other important nutrients, especially P. In fact, if the N:P ratios of receiving waters are high (i.e., higher than 16:1 in atomic units; Redfield 1934), the impacts of enhanced N-export may be low because P will be the principal element limiting freshwater production.

It is likely that N:P ratios in aquatic ecosystems will vary along a predictable trajectory across the phases of development presented above. Background N:P ratios tend to be very high in waters exported from undisturbed tropical terrains (Bruijnzeel 1991; Williams & Melack 1997), often higher than 100:1 (as atoms; Table 1). In minimally impacted tropical freshwaters, denitrification is quite rapid (Salati et al. 1982; Saijo et al. 1997) and warm, anoxic waters keep P mobile (Furch & Junk 1993), so N:P ratios are typically quite low (Bonetto et al. 1994; Pedrozo & Bonetto 1989; Pedrozo et al. 1992; Rai & Hill 1980). In some tropical regions, however, oxic conditions and waters rich in iron and dissolved organic matter may reduce P concentrations keeping N:P high (Gonzalez et al. 1991; Tundisi et al. 1991).

As forests are cleared, the N:P in exported waters will decrease. Storm related erosion, common in deforested areas, yields significantly lower N:P due primarily to increased transport of particulate matter containing relatively high concentrations of P. Burning volatilizes more of the N contained in biomass than the P (Hölscher et al. 1997; Kauffmann et al. 1995). The N:P ratio should therefore decline steadily throughout the burning phase of deforestation (Figure 5) unless most volatilized N is redeposited locally. Agricultural development can have different impacts on N:P ratios, because grain cultivation (e.g., maize) often uses large nitrogen amendments (e.g., anhydrous ammonia), which leads to high N:P ratios of runoff, while animal husbandry leads to lower N:P ratios (e.g., <20) characteristic of effluents

Table 1. Data represent the N:P ratios (molar) found in various known aquatic nutrient sources or in water fluxes leaving different land-uses.

Source	N:P (atomic)	Reference
Unfertilized fields	547	Loehr 1974
Soils, medium fertility	166	Vollenweider 1968
Forested areas	157	Loehr 1974
Rural lands	135	Loehr 1974
Soils, fertile	74	Vollenweider 1968
Tropical forests	52	Loehr 1974
Agricultural watersheds	44	Uttormark et al. 1974
Forest export	38	Beaulac & Reckhow 1982
Mixed agriculture	35	Beaulac & Reckhow 1983
Sewage effluent	22	Golterman 1975
Cattle manure seepage	20	Loehr 1974
Non-row crop export	18	Beaulac & Reckhow 1982
Fertilizer, average	17	Turner & Rabalais 1991
Redfield ratio	16	Redfield 1934; Harris 1986
Feedlot runoff	14	Loehr 1974
Pastureland runoff	13	Beaulac & Reckhow 1982
Urban stormwater	13	Loehr 1974
Sewage	12	Vollenweider 1968
Urban runoff	11	Beaulac & Reckhow 1982
Pastureland runoff	11	Loehr 1974
Urban runoff	10	Uttormark et al. 1974
Row crop export	9	Beaulac & Reckhow 1982
Sewage	6	Vallentyne 1974
Septic tank effluent	6	Brandes et al. 1974
Bird feces	2	Portnoy 1990

from manures (Table 1). The N:P ratio should further be reduced as nutrient-enriched waters pass through impoundments, especially if they are deep and have long water-residence times, because denitrification occurs at the sediment-water boundary in reservoirs and soluble P is released in anoxic sediments. Some upward pressure may be put on N:P, however, if land clearing leads to drastically altered hydrology that flushes normally anoxic waters where denitrification was naturally common. Urban and industrial sewage has a low N:P ratio (Downing & McCauley 1992), so long-term trends in heavily impacted tropical regions should stabilize at low N:P ratios typical of urbanized conditions. Thus, not only should nutrient export change with land-use, but nutrient supply ratios are likely to decline as development of tropical watersheds proceeds.

Tropical freshwater responses to N- and P-loading

The relative importance of N or P as a limiting nutrient depends not only upon their ratio in the inputs, but on several internal processes (Howarth 1988). These internal processes include recycling within the water column, sedimentation, resuspension or release from the bottom, and nitrogen fixation. Hence, proper evaluation of freshwater responses to altered N-loading requires a comparative analysis of how these processes alter nutrient cycling and availability.

As indicated above, disturbance will likely lead to a decrease in ratios of available N and P, although assembling a data set that is geographically representative of loadings to tropical lakes and streams is problematic. Different N and P fractions (i.e. dissolved inorganic, total dissolved, dissolved organic, particulate, or total) have been measured in various systems, and it is unclear whether the measured fractions are available for biological uptake, or whether sampling or analytical problems have yielded a clear view of tropical nutrient fluxes. Phrasing the hypothesis as a comparison between tropical and temperate regions may also confound the analysis, because the variability arising from land-use patterns could override that arising from geographical differences. Hence, the data we have chosen to make such comparisons below are from the least disturbed catchments possible.

The N:P ratios of total N and P fluxes in rivers draining into the North Atlantic (Table 2) supports the hypothesis that tropical freshwaters should generally be more sensitive to N-loading than temperate ones because the N:P ratios in tropical water flowing from large tropical drainage basins are at, or below, the Redfield atomic ratio (ca. 16), while the ratios in water leaving large temperate American and European basins (even those in undeveloped areas) are generally above 16 (Søballe & Kimmel 1987). Another approach is to examine yields of N and P from different land-uses. Based on North American data, as the dominant land-use changes from forests or mixed agriculture to pastures, crops, and finally to urban uses, total N to total P export shifts from values above atomic N:P ratios of 16 to values near or below 16 (Table 1). Moreover, Billen et al. (1991) found that total dissolved nutrient concentrations increased and dissolved N:P ratios decreased in a set of temperate catchments with land-use ranging from forested to agricultural to urbanized. Further, Peierls et al. (1991) reported strong statistical evidence for increased nitrate concentrations in rivers worldwide as a function of

Table 2. Regional N to P atomic ratios (Howarth et al. 1996) expressed as ratios of total N fluxes to total P fluxes. Data are for river systems flowing into the North Atlantic Ocean.

River/ Region	N:P Ratio
North America	42
Central America, Caribbean and Orinoco basin	15.5
Amazon and Tocantins basin	4.4
Western Europe	22

population density, and Howarth (1998) found the same relationship for total N export from large temperate regions. Therefore, a tendency toward lower N to P ratios favoring N limitation could be attributed to increased agricultural development and urbanization independent of latitude.

Detailed analysis of the relative severity of N or P limitation in freshwaters requires N and P data on a wide range of tropical waters, although complete data are rarely available. Although not necessarily reflecting total N:P, more frequently available nitrate to orthophosphate molar ratios vary widely within the tropics (Table 3) with values for African rivers and streams below the molar Redfield ratio (16) and most values for South America above 16. The N:P ratios of fluvial fluxes of total dissolved N and P for the South and Central American waters in Table 4 are at or above the Redfield ratios. Waters of the Paraguay River (Pedrozo & Bonetto 1989), the Paraná flood plain (Pedrozo et al. 1992), some Amazonian lakes (Forsburg 1984; Rai & Hill 1980) have low N:P, while other Amazon lakes (Forsburg 1984) and some lakes in Brazil (Tundisi et al. 1991) and Venezuela (Gonzalez et al. 1991) have high N:P. It appears that there may be regional differences in the prevalence of N limitation among the freshwater ecosystems of the tropics.

There are few measurements of atmospheric deposition in the tropics spanning at least one year and including inorganic and organic fractions of N and P (e.g. Bootsma et al. 1996; Lesack & Melack 1991; Lewis 1981); such data are sparse even for temperate regions. Hence, it is not possible to assemble a representative set of data on external loading via atmospheric deposition appropriate for testing hypotheses about the relative importance of N or P in tropical systems. The few data available show molar N to P ratios for dissolved inorganic N:P or total dissolved N and P that range from mid 20s to almost 200. Dry deposition of P is likely to lower these ratios, assuming that the particulate P is available for algal uptake.

The extent to which the N:P ratio in external inputs determines the nutrient status of freshwater ecosystems depends partially on the relative magnitude

Table 3. N to P ratios of rivers in various tropical regions expressed as molar NO_3 to PO_4 concentrations. Ratios of dissolved inorganic nutrients are used here because data in total river N and P budgets are frequently unavailable in the tropics.

River/ Region	Ratio
Africa	
Uganda (Viner 1975)	
Mt. Elgon rivers	4
Karamoja rivers	2
Central Uganda	3
Semliki tributaries and Ituri forest streams	5
Kigezi highlands rivers	5
Ruwenzori mountains	13
Kenya (Melack & McIntrye 1991)	
Aberdare and Bahati rivers	8
Cherangani Hills rivers	2
Zambezi (Hall et al. 1976)	1
South America (Lewis et al. 1995)	
Orinoco basin	
Orinoco (lower)	19
Apure	7
Caura	63
Caroni	87
Amazon basin	
Amazon (Obidos)	14
Solimoes (Ica)	16
Japura	24
Jurua	21
Negro	30
Madeira	10
Trombetas	62
Xingu	57
Tapajos	100

Table 4. N to P ratios expressed as fluxes of total dissolved N to total dissolved P or as fluxes of NO_3-N plus dissolved organic N to PO_4-P plus dissolved organic P. These nutrient ratios are used here because data on total river N and P budgets are frequently unavailable in the tropics.

River/ Region	Ratio (molar)	Reference
Paraguay	51	Hamilton et al. (1997)
Amazon basin		
Ica	20	JR Richey (unpubl.)
Jutai	20	JR Richey (unpubl.)
Jurua	15	JR Richey (unpubl.)
Japura	18	JR Richey (unpubl.)
Purus	18	JR Richey (unpubl.)
Negro	15	JR Richey (unpubl.)
Madeira	15	JR Richey (unpubl.)
Mota Brook	194	Lesack (1993)
Mota Brook	46	Williams & Melack (1997)
Orinoco (lower)	31	Lewis & Saunders (1989)
Apure	15	Saunders & Lewis (1988)
Costa Rica		
Northwestern streams	42	Newbold et al. (1995)

of external inputs, internal recycling and vertical mixing and losses, and their respective N:P ratios. Such information is available for only a few tropical lakes. Two contrasting examples (Lake Malawi and Lake Calado, Brazil) help demonstrate the balance between external and internal nutrient supply.

Lake Malawi, one of the African Great Lakes, has a hydraulic flushing time of 750 years and is permanently anoxic below about 200 m (Hecky et al. 1996). The annual fluvial and atmospheric inputs to the mixed layer have N to P molar ratios of 23 and 43, respectively. About 80% of the N input is supplied by nitrogen fixation. Vertical mixing provides about 88% of the P, but essentially no N to the mixed layer. Fluvial outflow is small and has an N:P ratio of 30; loss to burial in sediments is unmeasured; an N:P ratio for annual burial of about 30 is estimated for neighboring Lake Tanganyika. Although the N:P ratios in fluvial and atmospheric inputs would indicate a P-limited mixed layer, the dominance of vertical mixing as the route of supply of phosphorus (N:P $\approx$ 0), requires a large N supply by nitrogen fixation to avoid severe N limitation. The presence of a continuous oxic-anoxic interface in the water column acts as a sink for dissolved inorganic N because of denitrification, but has little influence on the upward flux of P (Hecky et al. 1996).

The common occurrence of anoxic hypolimnia in tropical lakes may be an important factor leading to low N to P ratios associated with vertical mixing, which is often a major source of nutrients to the euphotic zone, and which may stimulate N fixation or lead to N limitation.

Lake Calado, a shallow floodplain lake bordering the Solimões River in the central Amazon basin, ranges in area from about 2 to 8 km^2 as a function of annual variation in river stage (Lesack & Melack 1995). Estimates of inputs, recycling and exports of N and P for Lake Calado have been published (Doyle & Fisher 1994; Fisher et al. 1991; Melack & Fisher 1990). Inputs to Lake Calado from rain, upland runoff, groundwater seepage, adjacent lakes, and riverine inflows had an aggregate molar N:P ratio of 32. Internal recycling by bacteria and zooplankton within the water column and by regeneration from sediments had molar N:P ratios of 4 and 7, respectively. Because the magnitude of internal recycling exceeded external supply, the overall N:P ratio for supply was 5. Losses via burial as sediments and outflow to the river (but not including gaseous losses of N) had an N:P ratio of 20. Nitrogen fixation provided about 8% of the total annual N inputs. While these annual N:P supply ratios are indicative of N limitation, experimental enrichments, and physiological assays detected weekly and seasonal differences in nutrient limitation due in part to the variable incursion of river water (Setaro & Melack 1984). Hence, short-term variations in the relative magnitude of various routes of nutrient supply with different N:P ratios can change the importance of N or P as limiting nutrients. The Lake Malawi and Calado data suggest that increases in N loading of freshwaters will have a proportionately greater impact on tropical lakes than temperate ones.

Some information on the relative probabilities of N- or P-limitation in tropical freshwaters can be gained by examining average total N and P concentrations in tropical waters. Although such data are relatively rare, a joint PAHO/WHO program created to predict responses of unimpacted tropical lakes to cultural eutrophication assembled N and P data on 24 warm, tropical lakes in Argentina, Brazil, Colombia, Ecuador, Mexico, Puerto Rico, and Venezuela (Salas & Martino 1991). When these data are compared with similar data collected in temperate lakes (Downing & McCauley 1992), it appears that they follow quantitatively different trends (Figure 7). Tropical lakes in the Americas are generally very rich in nutrients and therefore have N:P ratios that cluster closer to N:P = 16, but are not dissimilar from rich temperate lakes. A greater proportion of tropical lakes (29%) than temperate (3%) have N:P $<$ 16 (Downing & McCauley 1992). This suggests that denitrification in tropical ponds, lakes, and streams must be very rapid because water leaving tropical forests has high N:P ratios (Figure 7; Bruijnzeel 1991; Williams & Melack 1997).

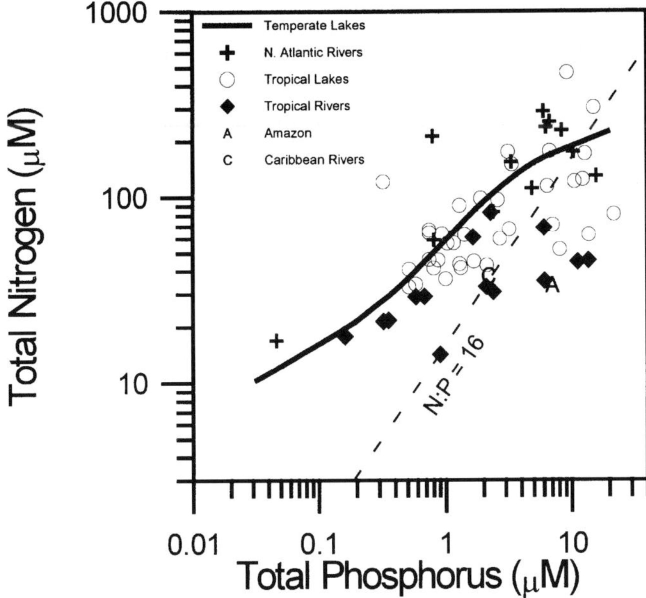

Figure 7. Relationship between the nitrogen and phosphorus concentration in temperate and tropical freshwater ecosystems. The solid line shows data on temperate lake nutrients shows the LOWESS fitted general trend derived from several hundred sets of measurements on average temperature lake nutrients (Downing & McCauley 1992). Tropical lake data are from Salas & Martino (1991). Concentrations of N and P in tributaries to the North Atlantic, including Caribbean rivers ("C") and the Amazon & Tocantins Basins ("A") are flow weighted averages from Howarth et al. (1996). Other tropical river data are principally pristine South American rivers and streams (Lesack 1993; Lewis et al. 1995) and data from the Gambia River in Africa (Lesack 1984).

Because low ambient N:P ratios are indicative of N limitation (Downing & McCauley 1992; Elser et al. 1990), it is likely that N is more frequently limiting to freshwater production in tropical lakes despite the high N:P in nutrient supplies from undisturbed forested catchments. Likewise, undisturbed tropical streams and rivers have lower concentrations of N for a given concentration of P than do tropical lakes or temperate lakes and rivers (Figure 7). This suggests that N is lost from aquatic systems more rapidly in the tropics than the temperate zone (e.g. through denitrification), and phosphorus erosion from tropical landscapes and phosphorus mobilization in aquatic environments are more rapid. Whatever the pathway, it is clear that nitrogen is frequently limiting to production in tropical freshwaters (Saijo et al. 1997; Furch & Junk 1993; Forsburg 1984; Henry et al. 1984).

In summary, it is difficult to predict with certainty how tropical freshwater systems will respond to altered nutrient supplies and ratios because of the relative paucity of data. It is especially important to make measurements of organic N and P fractions in atmospheric deposition, and to conduct studies that include recycling, vertical mixing and losses as well as inputs. Based on the available data on N:P ratios in fluvial inputs and lakes, however, it seems likely that N will be more frequently limiting to production in tropical freshwater systems than in the temperate zone. The common occurrence of an oxic-anoxic interface in the water column of tropical lakes may increase the likelihood of low N:P ratios, stimulating nitrogen fixation or leading to intensified nitrogen limitation.

Ecological effects of N/P alterations on tropical aquatic communities

The evidence presented above suggests that the probable sequence of disturbance will lead to N and P enrichment and progressive declines in the N:P of nutrient supplies. The coupling of the major biogeochemical cycles is critically important for evaluating the ecological impact of changes in them (Elser in review; Howarth 1988; Howarth et al. 1996; Sterner et al. 1997; Wollast et al. 1993). Resource ratio competition theory (Tilman 1982) can help develop broad predictions about the impact of increased N loading and stoichiometric shifts in supply with tropical disturbance.

To illustrate stoichiometric changes in tropical aquatic ecosystems, “Zero net growth isoclines” (combinations of N and P supply rates where population growth rate exactly balances losses such that net growth is zero) for nine hypothetical algal species are shown in Figure 8. The competitive abilities of various algal taxa usually follow a normal distribution (Andersen 1997), and thus in most situations the greatest diversity of algal taxa will be supported at intermediate N:P supply ratios. The species at the upper left of the figure are good competitors for P, but relatively poor competitors for N. The species to the right are good competitors for N but poor competitors for P. The “ultimate” good competitor for N is an N-fixing species (e.g. species 8 and 9) that might be abundant under the relatively N-limited conditions in the tropics. If much N, but little P, is supplied, good P-competitors will dominate. If much P, but little N, is supplied, as will probably be the case under increased disturbance in the tropics, good N-competitors (e.g., species 7 and 8) will reduce N levels below those at which other taxa can survive.

Resource ratio theory predicts different community responses of temperate and tropical systems to increased loading. Temperate lakes are usually P-limited with naturally high N:P loading ratios (Region A, Figure 8), so increased N loading accompanied by a proportional increase in P inputs (arrow I, Figure 8) has had significant yet modest effects on community structure. In N-limited (low N:P) waters (scenario B, Figure 8) or those

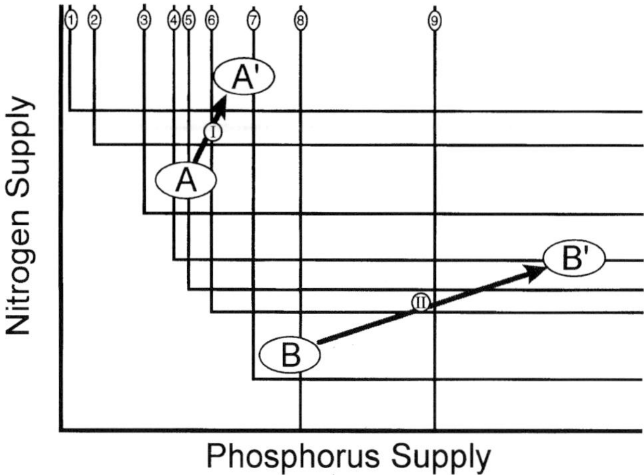

Figure 8. A generalized set of "zero net growth isoclines" ('ZNGI's) for 9 hypothetical algal species in which there is a trade-off in competitive abilities for N and P. For a given species, the ZNGI indicates the set of minimal supply rates for N and P at which net growth rate falls to zero for a given total loss rate. Species 1 and 2 are particularly good competitors for P but have elevated requirements for N supply. Conversely, species 8 and 9 represent N-fixing taxa with high minimal P requirements. The trajectories indicated by arrows I and II show scenarios of algal community response to perturbations of N supply under contrasting ecological conditions, representing changes observed in temperate freshwaters (I) and predicted for tropical freshwaters (II).

where N and P are nearly co-limiting, such as those that are common among oligotrophic lakes in the western U.S. (Horne & Galat 1985; Morris & Lewis 1988) and among many lakes in the tropics (Figure 7.; Lewis 1996; Setaro & Melack 1984; Wurtsbaugh et al. 1985), increased N loading that is coupled to increased P loading (N:P load decreasing; arrow II, Figure 8) would have relatively large effects on algal community structure, simplifying the community to those very few species adapted to extremely low N:P supply ratios. Given the predominantly low N:P of runoff under anthropogenic influence (Table 1), increasing disturbance in the tropics will lead to disproportionate enrichment of P relative to N, resulting in decreasing N:P ratios (Figure 5). Nitrogen limitation is therefore likely to become progressively more severe as disturbance and development proceed, leading to an even greater decrease in algal diversity than that seen in culturally eutrophied temperate freshwaters.

Discrepancies between the highly variable elemental composition of phytoplankton and the rigidly regulated composition of consumers (Andersen & Hessen 1991; Elser et al. 1996; Elser et al. 1998; Sterner et al. 1993; Sterner & Hessen 1994) will likely destabilize tropical freshwaters under increasing landscape disturbance. The growth and reproduction of consumers can be limited by the nutrient content of its food (Sterner 1994), creating trophic bottlenecks (Elser et al. 1998; Sterner et al. 1997) that limit the rates and ratios at which consumers cycle nutrients (Elser et al. 1988; Olsen et al. 1986; Sterner 1990; Urabe 1995). Models (i.e., Andersen 1993, 1997) predict that if N loading is coupled to increased P loading as expected in tropical freshwaters, trophic destabilization will result from tropical eutrophication. Because major biomolecules (e.g. proteins vs. nucleic acids) differ little in nitrogen content (Elser et al. 1996) but greatly in phosphorus content, most consumer taxa should have similar nutritional requirements for N but broadly variable P requirements. Worsening food quality resulting from N limitation may eventually drive food C:N ratios to a critically high level where the majority of grazer taxa suffer nutritional imbalances. Under increasing algal N limitation such as that expected in the tropics, impacts on food webs may potentially disruptive to ecosystem function. Because tropical freshwater systems are currently frequently N-limited, it is likely that ecosystem-level impacts of land-use change will be more severe than those seen in the temperate zone.

Altered nutrient export to estuaries

Evidence has been presented above that tropical freshwaters supply more nutrient flux than undisturbed rivers of the temperate zone (Figures 1 and 7), that the nitrogen contained in this flux is roughly 35% dissolved inorganic N (Lewis et al., this volume), and that tropical headwater streams have lower N:P ratios than their counterparts in the temperate zone (Table 2, Figure 7). Future land-use change in the tropics will be likely to greatly alter N flux to tropical estuaries (Figure 3); this N-flux will be increasingly dominated by dissolved inorganic forms of N (Figure 4); and nutrients may be increasingly supplied at N:P below the Redfield ratio (Table 1, Figure 3). These observations combine to suggest that tropical estuaries of the future will experience nutrient supplies that are much higher than current concentrations, and that these supplies will have decreasing N:P ratios leading to eutrophication and stronger nitrogen limitation. Increased nutrient loading to estuaries will probably accentuate nitrogen limitation.

Shifts in N and P limitation in tropical coastal ecosystems

The often high levels of nutrient loading and the low N:P in inputs to tropical coastal systems from rivers suggest that marine waters near river outflows must frequently be N-limited. Suspended silt loads carried by tropical rivers, however, block light and probably also result in light-limited production of benthic and planktonic systems. Nutrient inputs and factors limiting primary productivity in tropical coastal waters will therefore vary according to a continuum of geomorphological settings that differ in nutrient concentrations and light availability from deltaic, estuaries and lagoons, to reef and planktonic environments. The amounts of solar, chemical and mechanical energy together with regional geomorphology must establish the template from which site-specific patterns of primary productivity will develop. Together, these geophysical factors control the distribution of resources (such as light and nutrients) as well as establishing levels of stress (temperature and salinity) that largely determine the net productivity of aquatic ecosystems. These characteristics will strongly influence the responses of marine ecosystems to increased nutrient flux.

Most classification systems of marine ecology have accounted for the importance of the physical nature of coastal environments to the types of communities that develop (Hedgpeth 1957; Odum et al. 1972). Geomorphic and geophysical characteristics of coastal zones, along with regional climate, have been recognized as important constraints on the structure of tropical coastal ecosystems (Yanez-Arancibia 1987) including plant formations, such as mangroves (Blasco 1984; Thom 1982; Twilley 1995). In addition, the relative contributions of primary producers to total energy flow of coastal systems can be associated with regional patterns of geomorphology and hydrography, particularly river inflow (Mann 1975; Welsh et al. 1982). For example, the ratio of wetland area to the area of coastal waters has been suggested as important to the contribution of coastal wetlands to the secondary productivity of estuaries (Mann 1975; Rojas Galaviz et al. 1992; Welsh et al. 1982). Landform characteristics of a coastal region and environmental processes such as river discharge, tides, wind and waves control the types of primary producers, their relative spatial distribution and abundance, as well as rates of primary productivity. Classes of environmental settings based on energy spectrum and source of sediments have been used to generalize patterns of coastal geomorphology including reefs, lagoons, estuaries (drowned river valleys), and deltas (Figure 9). The response of coastal ecosystems to increases in N will likely be associated with these different patterns of geophysical forces and geomorphological processes.

The susceptibility of tropical coastal systems to eutrophication will vary with the geomorphological setting (Figure 9). Conceptual templates of coastal systems that integrate geomorphology and water residence time (geophysical processes) with biogeochemical processes allow understanding of the ways in which coastal ecosystems will respond to nutrient inputs. Variation of available light and nutrients are two key factors limiting primary productivity among the geomorphological types. Reefs are among the marine environ-

rapidly than N loading, as expected (see above and Figure 5), it is likely that P limitation will be replaced by N limitation, even in calcareous tropical marine ecosystems. There is a critical need for information on the biogeochemistry of C, N, and P in coastal ecosystems of the tropics, particularly in reaction to shifts in N:P ratios as a result of internal nutrient regeneration processes (see similar needs in lakes, above). The conceptual framework proposed here suggests that these patterns of biogeochemistry and susceptibility to N loading will vary depending on the geophysical and geomorphological characteristics of the coastal landscape.

Evidence of the effects of anthropogenic nutrients on coastal tropical ecosystems

Tropical coastal planktonic communities are most frequently nitrogen or light limited, and the impact of natural river exported nutrients (low in N:P) will eventually extend far beyond the river mouth. The seasonal influence of tropical rivers such as the Orinoco and Amazon has been shown to extend throughout the Eastern Caribbean Basin (Yoshioka et al. 1985). Recent evidence suggests that tropical river-borne nutrients are carried great distances because they are complexed with dissolved organic matter, and these complexes only slowly degrade by UV and bacterial activity (Bonilla et al. 1993; Carlsson et al. 1993; Moran & Hodson 1994). Although there is some controversy concerning the relative roles of N and P in limiting marine plankton production (Downing 1997; Downing et al. 1999; Howarth 1988; Smith 1984; Vitousek & Howarth 1991), the predominance of nitrogen fixing cyanobacteria in nutrient depleted oceanic waters (Carpenter 1983; Margalef 1971) and the preferential uptake of nitrogen by phytoplankton when nutrient limitation is relieved through upwelling (Corredor 1979) suggest that N is the most frequently limiting nutrient in the Caribbean and tropical Atlantic planktonic systems. Further evidence of current N limitation in the tropics is supplied by the high rates of N limitation seen in coastal waters (Diaz et al. 1990), and the strong responses of cyanobacteria to inorganic N enrichment (Corredor et al. 1994).

Human activities in the tropics have already begun to cause large imbalances in the marine coastal nitrogen cycle. Anthropogenic nutrient inputs due to increased fertilizer use, land development, soil erosion and sewage discharges have brought about rapid changes in algal communities. Filamentous cyanobacteria and chlorophytes have begun to dominate the phytobenthos in many places. The filamentous cyanophyte *Microcoleus lyngbyaceus* has now formed monospecific mats over the shallow benthos around Puerto Rico, in response to anthropogenic eutrophication (Corredor et al. 1994; Diaz et al. 1990). These mats have displaced the normal benthic community and depleted dissolved oxygen. Fleshy green algae can rapidly overgrow corals in response to anthropogenic nutrient inputs, as shown for Hawaiian reefs (Smith 1984). While excessive proliferation of these species has not been documented in marine waters of the American tropics, other algae (e.g. the calcareous green alga *Halimeda*) have shown rapid growth (Corredor, unpublished). Increased inputs of nutrients to tropical systems are already leading to large changes in these communities.

Although coastal denitrification significantly buffers the impact of increased nitrogen loading in the temperate zone (Seitzinger 1988), nitrification in tropical marine systems may be more severely constrained by relatively low oxygen concentrations of warm waters, especially those with active microbial floras typical of polluted coastal zones. Denitrification rates in mangrove ecosystems are controversial (Corredor & Morell 1994; Rivera-Monroy & Twilley 1996). Nitrogen delivered to the tropical marine ecosystem, therefore, will probably persist longer and thereby have a greater impact than in temperate marine ecosystems.

Overview of hypotheses and needs for future research

The preceding overview and comparisons suggest that differences in background nutrient environments between tropical and temperate systems will lead to greater impacts of human population growth and agricultural and industrial development on tropical aquatic ecosystems. Tropical freshwater systems are more frequently N-limited than temperate zones, while tropical marine systems may be more frequently P-limited than comparable areas in temperate zones. This means that disturbances to pristine tropical lands will likely lead to greatly increased primary production in freshwaters, an expansion of the high production, N and light-limited zones around river deltas, and a switch from P- to N-limitation with large changes in the community composition of fragile mangrove and reef ecosystems.

This review was assembled using as much empirical data on tropical nutrient cycles as we could find, but many facets of this issue will only be resolved by filling in critical gaps in existing knowledge. Although it is clear from existing evidence that N loss rates are higher from undisturbed tropical ecosystems than their temperate counterparts, it is unclear why these N-export rates are so high or how these N-loss mechanisms will react to altered land-use. Under pristine conditions, even modeled atmospheric N-deposition is calculated to be higher in the tropics than temperate zones, but we currently have little solid data on how atmospheric N-deposition might react to disturbance or how expected increases in NO_x:NH_y will impact denitrification or river transport of atmospheric N. Research on the mechanisms of nutrient transport and atmospheric deposition in the tropics is vital to understanding the probable impacts of land-use disturbance.

We have used a conceptual model built from tropical and temperate geochemical experience to project a trajectory of change in aquatic N concentrations and N:P stoichiometry as land-use is altered in the tropics. Although we think it likely that tropical ecosystems will follow this trajectory, the N-flux and N:P database is quite thin and needs field testing. Critical questions concern shielding tropical aquatic systems from excess nutrient, sediment, and nutrient loads through the use of riparian buffers and other management practices throughout the phases of deforestation, burning and agricultural development. The proper management of reservoirs and water-management projects offer a potentially important means of controlling nutrient loads, thus research is needed to examine characteristics of the growing number of tropical impoundments that might lead to more effective nutrient retention and lowered downstream impacts on freshwater and marine ecosystems. Research is needed to evaluate ways to buffer the major changes in nitrogen export that can be expected from large-scale disturbance and development of tropical lands.

We strongly emphasize the need for research that recognizes the impact of altered nutrient stoichiometry on tropical aquatic ecosystem function. Although it appears that N limitation of freshwater systems is more prevalent in the tropics than temperate zones, research on the prevalence and mechanism of N limitation in tropical ecosystems is needed to understand how freshwater production and community structure will be altered by land-use changes. The database on N and P cycling in tropical freshwater lakes, ponds and rivers is particularly incomplete, and we need a fuller understanding of the mechanisms coupling or uncoupling the fluxes of N, P and other major nutrients from disturbed tropical ecosystems. Relationships between tropical aquatic community structure and nutrient stoichiometry must be worked out, especially to predict how noxious algal species are impacted by altered stoichiometry and nutrient supply. The potential destabilization of trophic interactions by altered nutrient ratios suggests that research is needed to determine the optimal N and P needs of key components of tropical aquatic food webs.

The tropics contain a large variety of marine ecosystems that vary in nutrient limitation from P-limited mangroves and reefs to N-limited coastal zones to deltaic receiving waters limited by light penetration through turbid waters. Prediction of the impact of altered nutrient flux requires a better understanding of how light, N and P limitation combine to determine rates of production and community structure in tropical marine littoral, benthic, and planktonic ecosystems. Disruption of land-use patterns will clearly alter nutrient availability to the marine environment, but it is important to determine how light and nutrients might interact to alter large-scale patterns in marine production. Impacts of changing ratios of nutrient supply on calcareous tropical marine communities (e.g. reefs) are particularly poorly understood, thus prediction of tropical marine impacts requires knowledge of the mechanisms through which nutrient stoichiometry influences marine nutrient limitation.

Land-use changes have greatly altered temperate nutrient cycles, and similar land-use changes are occurring at a rapid rate in the tropics. The current base of biogeochemical data suggests that alterations in the N-cycle will have greater impacts on tropical ecosystems than those already seen in temperate zones. As tropical development proceeds, the research outlined above could go far toward protecting tropical aquatic ecosystems from the problems experienced in the temperate zone as a result of anthropogenic enhancement of nutrient flux.

References

Andersen T & Hessen DO (1991) Carbon, nitrogen, and phosphorus content of freshwater zooplankton. Limnol. Oceanogr. 36: 807–814

Andersen T (1993) Grazers as Sources and Sinks for Nutrients. PhD Dissertation, University of Oslo, Norway

Andersen T (1997) Pelagic Nutrient Cycles: Herbivores as Sources and Sinks. Springer-Verlag, Berlin, Germany

Arcifa MS, Carvalho MAJ, Gianesella-Galvao SMF, Shimizu GY, Froehlich CG & Castro RMC (1981a) Limnology of ten reservoirs in Southern Brazil. Verh. Internat. Verein. Limnol. 21: 1048–1053

Arcifa MS, Froehlich CG & Gianesell-Galvao SMF (1981b) Circulation patterns and their influence on physico-chemical and biological conditions in eight reservoirs in Southern Brazil. Verh. Internat. Verein. Limnol. 21: 1054–1059

Arheimer B & Witgren HB (1994) Modeling the effects of wetlands on regional nitrogen transport. Ambio 23: 378–286

Asner GP, Seastedt TR & Townsend AR (1997) The decoupling of terrestrial carbon and nitrogen cycles. BioScience 47: 226–234

Bailey C, Jentoft S & Sinclair P (1996) Aquacultural Development: Social Dimensions of an Emerging Industry. Westview Press, Boulder, CO, U.S.A.

Beaulac MN & Reckhow KH (1982) An examination of land-use-nutrient export relationships. Water Resources Bulletin 18: 1013–1024

Behrens CA, Baksh MG & Mothes M (1994) A regional analysis of Barí land use intensification and its impact on landscape heterogeneity. Human Ecology 22: 279–316

Billen G, Lancelot C & Meybeck M (1991) N, P, and Si retention along the aquatic continuum from land to ocean. In: Mantoura RFC, Martin JM & Wollast R (Eds) Ocean Margin Processes in Global Change (pp 19–44). Wiley & Sons, Chichester

Blasco F (1984) Climatic factors and the biology of mangrove plants. In: Snedaker SC & Snedaker JG (Eds) The Mangrove Ecosystem: Research Methods (pp 18–35). UNESCO, Paris, France

Bonetto C, De Cabo L, Gabillone N, Vinocur A, Donadelli J & Unrein F (1994) Nutrient dynamics in the deltaic floodplain of the Lower Paraná River. Arch. Hydrobiol. 131: 277–295

Bonilla J, Senior W, Bugden J, Zafiriou O & Jones R (1993) Seasonal distribution of nutrients and primary productivity on the eastern continental shelf of Venezuela as influenced by the Orinoco River. J. Geophys. Res. 98: 2245–2257

Bootsma HA, Bootsma MJ & Hecky RE (1996) The chemical composition of precipitation and its significance to the nutrient budget of Lake Malawi. In: Johnson TC & Odada EO (Eds) Limnology, Climatology and Paleoclimatology of the East African Lakes (pp 251–265). Gordon & Breach, Amsterdam, The Netherlands

Brandes M, Chowdhry NA & Cheng WW (1974) Experimental study on removal of pollutants from domestic sewage by underdrained soil filters. In: American Society of Agricultural Engineers, Proceedings of the National Home Sewage Disposal Symposium (pp 29–36)

Brodizio ES, Moran EF, Mausel P & Wu Y (1994) Land use change in the Amazon estuary: patterns of Caboclo settlement and landscape management. Human Ecology 22: 249–278

Bruijnzeel LA (1990) Hydrology of Moist Tropical Forests and Effects of Conversion: A State of Knowledge Review. Humid Tropics Program, UNESCO Int. Hydrological Program

Bruijnzeel LA (1991) Nutrient input-output budgets and tropical forest ecosystems: A review. J. Tropical Ecology 7: 1–24

Bruijnzeel LA (1996) Predicting the hydrological impacts of land cover transformation in the humid tropics: the need for integrated research. In: Gash JHC, Nobre CA, Roberts JM & Victoria RL (Eds) Amazonian Deforestation and Climate (pp 15–55). John Wiley & Sons, New York, NY, U.S.A.

Carlsson P, Segatto AZ & Granéli E (1993) Nitrogen bound to humic matter of terrestrial origin – a nitrogen pool for coastal phytoplankton? Marine Ecology Progress Series 97: 105–116

Carpenter EJ (1983) Nitrogen fixation by marine Oscillatoria (Trichodesmium) in the world's ocean. In: Carpenter EJ & Capone DG (Eds) Nitrogen in the Marine Environment. Academic Press, New York, NY, U.S.A.

Chowdhury MK & Rosario EL (1993) Nitrogen utilization efficiency as affected by component populations in maize-mungbean intercropping. Trop. Agric. 70: 199–204

Cole JJ, Peirls BL, Caraco NF & Pace ML (1993) Nitrogen loading of rivers as a human driven process. In: McDonnell MJ & Pickett STA (Eds) Humans as Components of Ecosystems: The Ecology of Subtle Human Effects and Populated Areas (pp 141–157). Springer-Verlag, New York, NY, U.S.A.

Cooke GD, Welch EB, Peterson SA & Newroth PR (1993) Restoration and Management of Lakes and Reservoirs, 2nd Edn. Lewis Publishers, Boca Raton, FL, U.S.A.

Corredor JE (1979) Phytoplankton response to low-level nutrient enrichment through upwelling in the Colombia Caribbean Basin. Deep-Sea Research 26A: 731–741

Corredor JE & Morell JM (1989) Nitrate depuration of secondary sewage effluents in mangrove sediments. Estuaries 17: 295–300

Corredor JE, Morell JM & Díaz MR (1994) Environmental degradation, nitrogen dynamics and proliferation of the filamentous cyanophyte Microcoleus lyngbyaceus in nearshore Caribbean waters. In: Amato E (Ed.) Mediterraneo e Caraibe due Mari in Pericolo? Sversamenti Accidentali di Idrocarburi ed Emegenze Causate Dalle Alghe. ICRAM/IFREMER. Atti Convegno Internatzionale, Genova

Day JW Jr & Kemp P (1985) Long-term impacts of agricultural runoff in a Louisiana swamp forest. In: Godfrey PJ, Haynor ER, Pelczarski S & Benforado J (Eds) Ecological Considerations in Wetlands Treatment of Municipal Wastewaters (pp 317–326). Van Nostrand, New York, U.S.A.

Deare FM, Ahmad N & Ferguson TU (1995) Downward movement of nitrate and ammonium nitrogen in a flatland ultisol. Fertilizer Research 42: 175–184

D'Elia CF & Wiebe WJ (1990) Biogeochemical nutrient cycles in coral-reef ecosystems. In: Dubinsky Z (Ed.) Ecosystems of the world 25, Coral reefs (pp 49–74). Elsevier, Amsterdam, The Netherlands

Díaz MR, Corredor JE & Morell JM (1990) Inorganic nitrogen uptake by Microcoleus lyngbyaceus mat communities in a semi-eutrophic marine community. Limnology & Oceanography 35: 1788–1795

Diaz RJ & Rosenberg R (1995) Marine benthic hypoxia: a review of its ecological effects and the behavioral responses of benthic macrofauna. Oceanography and Marine Biology: An Annual Review 33: 245–303

Diaz RJ, Neubauer RJ, Schaffner LC, Pihl L & Baden SP (1992) Continuous monitoring of dissolved oxygen in an estuary experiencing periodic hypoxia and the effect of hypoxia on macrobenthos and fish. Science of the Total Environment, Supplement 1992: 1055–1068

Downing, JA (1997) Marine nitrogen: Phosphorus stoichiometry and the global N:P cycle. Biogeochemistry 37: 237–252

Downing JA & McCauley E (1992) The nitrogen:phosphorus relationship in lakes. Limnol. Oceanogr. 37: 936–945

Downing JA, Osenberg CW & Sarnelle O (1999) Meta-analysis of marine nutrient-enrichment experiments: variation in the magnitude of nutrient limitation. Ecology 80(4): (in press)

Doyle RD & Fisher TR (1994) Nitrogen fixation by periphyton and plankton on the Amazon floodplain at Lake Calado. Biogeochemistry 26: 41–66

Eastwood DA & Pollard HJ (1992) Amazonian colonization in eastern Ecuador: land use conflicts in a planning vacuum. Singapore J. Tropical Geography 13: 103–113

Elser JJ, Chrzanowski TH, Sterner RW & Mills KH (1998) Stoichiometric constraints on food-web dynamics: a whole-lake experiment on the Canadian Shield. Ecosystems: in press

Elser JJ, Dobberfuhl D, MacKay NA & Schampel JH (1996) Organism size, life history, and N:P stoichiometry: towards a unified view of cellular and ecosystem processes. Bioscience 46: 674–684

Elser JJ, Elser MM, MacKay NA & Carpenter SR (1988) Zooplankton-mediated transitions between N and P-limited algal growth. Limnol. Oceanogr. 33: 1–14

Elser JJ, Marzolf ER & Goldman CR. (1990) Phosphorus and nitrogen limitation of phytoplankton in freshwaters of North America: A review and critique of experimental enrichments. Can. J. Fisheries and Aquatic Sci. 47: 1468–1477

Feller IC (1995) Effects of nutrient enrichment on growth and herbivory of dwarf red mangrove (Rhizophora mangle). Ecological Monographs 65: 477–505

Fernandez ECM, Biot Y, Castilla C, Canto AC, Matos JC, Garcia S, Perin R & Wanderli E (1997) The impact of selective logging and forest conversion for subsistence agriculture and pastures on terrestrial nutrient dynamics in the Amazon. Ciencia e Cultura, in press

Fisher TR, Lesack LFW & Smith LK (1991) Input, recycling, and export of N and P on the Amazon floodplain at Lake Calado. In: Tiessen H, Lopez-Hernandez D & Salcedo IH (Eds), Phosphorus Cycles in Terrestrial and Aquatic Ecosystems. Regional Workshop 3: South and Central America (pp 34–53). University of Saskatchewan Press, Regina

Fleisher S & Stibe L (1991) Drainage basin management – reducing river transported nitrogen. Verh. Internat. Verein. Limnol. 24: 1753–1755

Forsberg BR (1984) Nutrient processing in Amazon floodplain Lakes. Verh. Internat. Verein. Limnol. 22: 1294–1298

Fourqurean JW, Jones RD & Zieman JC (1993) Processes influencing water column nutrient characteristics and phosphorus limitation of phytoplankton biomass in Florida Bay, FL, U.S.A.: Inferences from spatial distributions. Estuarine Coastal and Shelf Science 36: 295–314

Furch K & Junk WJ (1993) Seasonal nutrient dynamics in an Amazonian floodplain lake. Arch. Hydrobiol. 128: 277–285

Galloway JN (1989) Atmospheric acidificatioN:Projections for the future. Ambio 16: 161–166

Galloway JN, Levy H & Kasibhatia PS (1994) Year 2020: Consequences of population growth and development on deposition of oxidized nitrogen. Ambio 23: 120–123

Garrity DP & Agustin DP (1995) Historical land use evolution in a tropical acid upland agroecosystem. Agriculture, Ecosystems and Environment 53: 83–95

Golterman HL (1975) Physiological Limnology: An Approach to the Physiology of Lake Ecosystems. Elsevier, Amsterdam, The Netherlands

Gonzalez E, Paolini J & Infante A (1991) Water chemistry, physical features and primary production of phytoplankton in a tropical blackwater reservoir (Embalse de Guri, Venezuela). Verh. Internat. Verein. Limnol. 24: 1477–1481

Hall A, Davies BR & Valente I (1976) Cabora Bassa: Some preliminary physicochemical and zooplankton pre-impoundment results. Hydrobiologia 50: 17–25

Hall CAS & Hall MHP (1993) The efficiency of land and energy use in tropical economies and agriculture. Agriculture, Ecosystems and Environment 46: 1–30

Hamilton SK, Sippel SJ, Calheiros DF & Melack JM (1997) An anoxic event and other biogeochemical effects of the Pantanal wetland on the Paraguay River. Limnology and Oceanography 42: 257–272

Harris GP (1986) Phytoplankton ecology: structure, function and fluctuations. Chapman & Hall

Hecky RE, Bootsma HA, Mugiddeand RM & Bugenyi FWB (1996) Phosphorus pumps, nitrogen sinks, and silicon drains: plumbing nutrients in the African Great Lakes. In: Johnson TC & Odada EO (Eds) Limnology, Climatology and Paleoclimatology of the East African Lakes (pp 205–224). Gordon and Breach, Amsterdam, The Netherlands

Hedgpeth JW (1957) Classification of marine environments. Geological Society of America, Memoir 67, Volume 1: 17–28

Hedin LO, Armesto JJ, & Johnson AH (1995) Patterns of nutrient loss from unpolluted, old-growth temperate forests: evaluation of biogeochemical theory. Ecol. 76: 493–509

Henry R, Tundisi JG & Curi PR (1984) Effects of phosphorus and nitrogen enrichment on the phytoplankton in a tropical reservoir (Lobo Reservoir, Brazil). Hydrobiologia 118: 177–185

Hölscher D, Möller RF, Denich M & Fölster H (1997) Nutrient input-output budget of shifting agriculture in Eastern Amazonia. Nutrient Cycling in Agroecosystems 47: 49–57

Horne AJ & Galat DL (1985) Nitrogen fixation in an oligotrophic, saline desert lake: Pyramid Lake, Nevada. Limnol. Oceanogr. 30: 1229–1239

Houghton RA (1994) The worldwide extent of land-use change. BioScience 44: 305–313

Howarth RW (1988) Nutrient limitation of net primary productivity in marine ecosystems. Annual Review of Ecology and Systematics 19: 89–110

Howarth RW (1998) An assessment of human influences on fluxes of nitrogen from the terrestrial landscape to the estuaries and continental shelves of the North Atlantic Ocean. Nutrient Cycling in Agroecosystems 52: 213–223

Howarth RW, Billen G, Swaney D, Townsend A, Jaworski N, Lajtha K, Downing JA, Elmgren R, Caraco N, Jordan T, Berendse F, Freney J, Kudeyarov V, Murdoch P, Zhao-Liang Z (1996) Regional nitrogen budgets and riverine N & P fluxes for the drainages to the North Atlantic Ocean: Natural and human influences. Biogeochemistry 35: 75–139

Jaworski NA, Howarth RW & Hetling LJ (1997) Atmospheric deposition of nitrogen oxides onto the landscape contributes to coastal eutrophication in the northeast United States. Environ. Sci. Technol. 31: 1995–2004

Juo ASR & Manu A (1996) Chemical dynamics in slash-and-burn agriculture. Agriculture, Ecosystems and Environment 58: 49–60

Justić D, Rabalais NN, Turner RE & Wiseman WJ Jr (1993) Seasonal coupling between riverborne nutrients, net productivity and hypoxia. Marine Pollution Bulletin 26: 184–189

Kauffman JB, Cummings DL, Ward DE & Babbitt R. (1995) Fire in the Brazilian Amazon: 1. Biomass, nutrient pools and losses in slashed primary forests. Oecologia. 104: 397–408

Keftasa D (1994) Effects of harvest management and nitrogen application on yield and nutritional value of Rhodes grass and lucerne in pure stands and mixtures. Trop. Agric. 71: 88–94

Kühne RF (1993) Wasser- und Nährstoffhauhalt in Mais-Maniok-Anbausgstemen mit und ohne Integration von Alleekultuven ("Alley cropping") in Süd-Benin. Honen Heimer Bodenkundliche Hefte 13: 1–244

Lesack LFW, Hecky RE & Melack JM (1984) Transport of carbon, nitrogen phosphorus and major solutes in the Gambia River, West Africa. Limnology & Oceaonography 29: 816–830

Lesack LFW & Melack JM (1991) The deposition, composition, and potential sources of major ionic solutes in the central Amazon basin. Water Resources Research 27: 2953–2977

Lesack LFW & Melack JM (1995) Flooding hydrology a mixture dynamics of lake water derived from mutiple sources in an Amazon floodplain lake. Water Resources Research 31: 329–341

Lesack LFW & Melack JM (1996) Mass balance of major solutes in a rainforest catchment in the Central Amazon: implications for nutrient budgets in tropical rainforests. Biogeochem. 32: 115–142

Lesack LFW (1993) Export of nutrients and major ionic solutes from a rain forest catchment in the central Amazon basin. Water Resources Research 29: 743–758

Lewis WM Jr (1981) Precipitation chemistry and nutrient loading by precipitation in a tropical watershed. Water Resources Research 17: 169–181

Lewis WM Jr (1996) Tropical lakes: how latitude makes a difference. In: Schiemer F & Boland KT (Eds) Perspectives in Tropical Limnology (pp 43–64). SPB Academic Publishing, Amsterdam, The Netherlands

Lewis WM Jr & Saunders JF (1989) Concentration and transport of dissolved and suspended substances in the Orinoco River. Biogeochemistry 7: 203–240

Lewis WM Jr, Hamilton SK & Saunders JF (1995) Rivers of northern South America. In: Cushing CE, Cummins KW & Minshall GW (Eds) River and stream ecosystems. Elsevier, Amsterdam

Loehr RC (1974) Characteristics and comparative magnitude of non-point sources. J. Water Pollution Control Federation 46: 1849–1872

Mann KH (1975) Relationship between morphometry and biological functioning in three coastal inlets of Nova Scotia. In: Cronin LE (Ed.) Estuarine Research Vol. 1 (pp 634–644). Academic Press, New York, U.S.A.

Margalef R (1971) The pelagic ecosystem of the Caribbean Sea. In: Symposium on Investigations and Resources of the Caribbean Sea and Adjacent Regions (pp 483–489). UNESCO

Matsumura-Tundisi T, Hino K & Claro SM (1991) Limnological studies at 23 reservoirs in southern part of Brazil. Verh. Internat. Verein. Limnol. 21: 1040–1047

McClain ME, Richey JE & Pimentel TP (1994) Groundwater nitrogen dynamics at the terrestrial-lotic interface of a small catchment in the Central Amazon Basin. Biogeochemistry 27: 113–127

McDowell WH & Asbury CE (1994) Export of carbon, nitrogen and major ions from three tropical montane watersheds. Limnology & Oceanography 39: 111–125

Melack JM & Fisher TR (1990) Comparative limnology of tropical floodplain lakes with an emphasis on the central Amazon. Acta Limnologica Brasil 3: 1–48

Melack JM & MacIntyre S (1991) Phosphorus concentrations, supply and limitation in tropical African lakes and rivers. In: Tiessen H & Frossard E (Eds) Phosphorus Cycles in Terrestrial and Aquatic Ecosystems, Regional Workshop 4: Africa (pp 1–18). Saskatchewan Institute of Pedology

Meybeck M (1982) Carbon, nitrogen, and phosphorus transport by world rivers. American J. Science 282: 401–450

Meybeck M (1998) Man and river interface: multiple impacts on water and particulates chemistry illustrated in the Seine river basin. Hydrobiologia 373: 1–17

Moran MA & Hodson RE (1994) Dissolved humic substances of vascular plant origin in a coastal marine environment. Limnol. Oceanogr. 39: 762–771

Morris DP & Lewis WM (1988) Phytoplankton nutrient limitation in Colorado mountain lakes. Freshwater Biol. 20: 315–327

Murdiyarso D & Wasriin UR (1995) Estimating land use change and carbon release from tropical forests conversion using remote sensing technique. J. Biogeography 22: 715–721

Neill C, Piccolo MC, Steudler PA, Melillo JM, Feigl BJ & Cerri CC (1995) Nitrogen dynamics in soils of forests and active pastures in the western Brazilian Amazon basin. Soil Biol. Biochem. 27: 1167–1175

Newbold JD, Sweeney BW, Jackson JK & Kaplan LA. (1995) Concentrations and export of solutes from six mountain streams in northwestern Costa Rica. J. North American Benthological Society 14: 21–37

Odum HT, Copeland BJ & McMahan E (1972) Coastal Ecosystems of the United States, Vols. 1–4. Conservation Foundation

Ojima DS, Galvin KA & Turner BL II (1994) The global impact of land-use change. BioScience 44: 300–304

Olsen Y, Jensen A, Reinertsen H, Borsheim KY, Heldal M & Langeland A (1986) Dependence of the rate of release of phosphorus by zooplankton on the P:C ratio in the food supply, as calculated by a recycling model. Limnol. Oceanogr. 31: 34–44

Pedrozo F & Bonetto C (1989) Influence of river regulation on nitrogen and phosphorus mass transport in a large South American river. Regulated Rivers: Research & Management: 4: 59–70

Pedrozo F, Diaz M & Bonetto C (1992) Nitrogen and phosphorus in the Parana River floodplain waterbodies. Arch. Hydrobiol./Suppl. 90: 171–185

Peierls BL, Caraco NF, Pace ML & Cole JJ. (1991) Human influence on river nitrogen. Nature 350: 386–387

Peterjohn WT & Correll DL (1984) Nutrient dynamics in an agricultural watershed: observations on the role of a riparian forest. Ecological Society of America 65: 1466–1475

Pinay G & Décamps H (1988) The role of riparian woods in regulating nitrogen fluxes between alluvial aquifer and surface water: A conceptual model. Regulated Rivers: Research and Management 2: 507–516

Portnoy JW (1990) Gull contribution of phosphorus and nitrogen to a Cape Cod kettle pond. Hydrobiologia 202: 61–69

Poss R & Sargoni H (1992) Leaching nitrate, calcium and magnesium under maize cultivation on anoxisol in Togo. Fert. Res. 33: 123–133

Prospero JM, Barrett K, Church T, Dentener F, Dunce RA, Galloway JN, Levy II H, Moody J & QuinnP (1996) Atmospheric deposition of nutrients to the North Atlantic Basin. Biogeochemistry 35: 27–73

Quirós R (1990) The Paraná River basin development and the changes in the Lower Basin fisheries. Interciencia 15: 442–451

Rai & Hill (1980)

Redfield AC (1934) On the proportions of organic derivatives in sea water and their relation to the composition of plankton. In: James Johnstone Memorial Volume (pp 169–192). Univ. Press, Liverpool, England

Rivera-Monroy VH & Twilley RR (1996) The relative role of denitrification an immobilization in the fate of inorganic nitrogen in mangrove sediments (Terminicos Lagoon, Mexico). Limnol. Oceanogr. 41: 284–296

Rojas Galaviz JL, Vera F, Yanez-Arancibia A & Day JW (1992) Estuarine primary producers: the Terminos Lagoon a case study. In: Seeliger U (Ed.) Coastal Plant Communities of Latin America (pp 141–154). Academic Press, NY, U.S.A.

Ross MR (1997) Fisheries Conservation and Management. Prentice Hall, Upper Saddle River, NJ, U.S.A.

Rudel T & Roper J (1996) Regional patterns and historical trends in tropical deforestation, 1976–1990: a qualitative comparative analysis. Ambio 25: 160–166

Saijo Y, Mitamura O, Hino K, Ikusima I, Tunidsi JG, Matsumura-Tundisi T, Sunaga T, Nakamoto N, Fukuhara H, Barbosa FAR, Henry R & Silva VP (1997) Physiochemical features of rivers and lakes in Pantanal wetland. Jpn. J. Limnol. 58: 69–82

Salas HJ & Martino P (1991) A simplified phosphorus trophic state model for warm-water tropical lakes. Water Research 3: 341–350

Salati E, Sylvester-Bradley R & Victoria RL (1982) Regional gains and losses of nitrogen in the Amazon Basin. Plant and Soil 67: 367–376

Salati E, Victoria RL, Martinelli LA & Richey JE (1991) Forests: their role in global change, with special reference to the Brazilian Amazon. In: Climate Change: Science, Impacts and Policy: Proceedings of the Second World Climate Conference (pp 391–395)

Saunders JF & Lewis WM Jr. (1988) Transport of phosphorus, nitrogen, and carbon by the Apure River,Venezuela. Biogeochemistry 5: 323–342

Seeliger U (1992) Coastal Plant Communities of Latin America. Academic Press, New York, U.S.A.

Seitzinger SP (1988) Denitrification in freshwater and coastal marine ecosystems: ecological and geochemical significance. Limnol. Oceanogr. 33: 702–724

Setaro FV & Melack JM (1984) Responses of phytoplankton to experimental nutrient enrichment in an amazon flood plain lake. Limnol. Oceanogr. 29: 972–984

Seyfried MS & Rao PSC (1991) Nutrient leaching loss from two contrasting cropping systems in the humid tropics. Trop. Agric. 68: 9–18

Short FT, Davis MW, Gibson RA & Zimmermann CF (1985) Evidence for phosphorus limitation in carbonate sediments of the seagrass syringodium filiforme. Estuarine, Coastal and Shelf Science 20: 419–430

Shukla J, Nobre C & Sellers P (1990) Amazon deforestation and climate change. Science 247: 1322–1325

Sjodin A, Lewis WM Jr & Saunders JF (1997) Denitrification as a component of the nitrogen budget for a large plains river. Biogeochem. 39: 327–342

Skole DL, Chomentowski WH, Salas WA & Nobre AD (1994) Physical and human dimensions of deforestation in Amazonia. BioScience 44: 314–324

Smith SV (1984) Phosphorus versus nitrogen limitation in the marine environment. Limnology and Oceanography 29: 1149–1160

Søballe DM & Kimmel BL. (1987) A large-scale comparison of factors influencing phytoplankton abundance in rivers, lakes, and impoundments. Ecology 68: 1943–1954

Soto D & Stackner J (1996) Oligotrophic lakes in southern Chile and in British Columbia: Basis for their resilience to present and future disturbances. In: Lawford, Alaback & Fuentes (Eds) High Latitude Rain Forest of the West Coast of the Americas. Climate, Hydrology, Ecology and Conservation (pp 266–280). Springer, NY, U.S.A.

Sterner RW (1990) The ratio of nitrogen to phosphorus resupplied by herbivores: zooplankton and the algal competitive arena. Am. Nat. 136: 209–229

Sterner RW & Hessen DO (1994) Algal nutrient limitation and the nutrition of aquatic herbivores. Ann. Rev. Ecol. Syst. 25: 1–29

Sterner RW, Elser JJ, Fee EJ, Guildford SJ & Chrzanowski TH (1997) The light:nutrient ratio in lakes: the balance of energy and materials affects ecosystem structure and function. Am. Nat. 150: 663–684

Sterner RW, Hagemeier DD, Smith WL & Smith RF (1993) Phytoplankton nutrient limitation and food quality for Daphnia. Limnol. Oceanogr. 38: 857–871

Sur HS, Mastana PS & Hadda MS (1992) Effect of rates and modes of mulch application on runoff, sediment and nitrogen loss on cropped and uncropped fields. Trop. Agric. 69: 319–322

Sussman RW, Green GM & Sussman LK (1994) Satellite Imagery, Human Ecology, Anthropology, and Deforestation in Madagascar. Human Ecology 22: 333–349

Teixeira C & Tundisi JG (1981) The effects of nitrogen and phosphorus enrichments on phytoplankton in the region of Ubatuba, Brazil. Bolm Inst. Oceanogr. 30: 77–86

Thom BG (1982) Mangrove ecology – a geomorphological perspective. In: Clough BF (Ed.) Mangrove Ecosystems in Australia (pp 3–17). Australian National University Press, Canberra

Tilman D (1982) Resource Competition and Community Structure. Princeton University Press, Princeton, NJ, U.S.A.

Tundisi JG (1981) Typology of reservoirs in southern Brazil. Verh. Internat. Verein. Limnol. 21: 1031–1039

Tundisi JG (1987) Local community involvement in environmental planning an management: the Lobo-Broa Reservoir case study. Regional Development Dialogue 8: 133–142

Tundisi JG (1994) Tropical South America: present and perspectives. In: R Margalef (Ed.) Limnology Now: A Paradigm of Planetary Problems (pp 353–424). Elsevier Science, The Netherlands

Tundisi JG & Matsumura-Tundisi T (1984) Comparative limnological studies at three lakes in tropical Brazil. Verh. Internat. Verein. Limnol. 22: 1310–1314

Tundisi JG, Matsumura-Tundisi T, Calijuri MC & Novo EML (1991) Comparative limnology of five reservoirs in the Middle Tietê River, S. Paulo State. Verh. Internat. Verein. Limnol. 24: 1489–1496

Turner RE & Rabalais NN (1991) Changes in Mississippi River water quality this century – implications for coastal food webs. BioScience 41: 140–147

Twilley RR (1995) Properties of mangrove ecosystems related to the energy signature of coastal environments. In: Hall CAS (Ed.) Maximum Power: The Ideas and Applications of H. T. Odum (pp 43–62). University Press of Colorado, Niwot, CO, U.S.A.

Twilley RR & Day JW (1998) The productivity and nutrient cycling of mangrove ecosystem. In: Yanez-Arancibia A & Lara-Dominguez AL (Eds) Mangrove Ecosystems in Tropical America: Structure, Function, and Management (pp 131–156). EPOMEX Scientific Series 3, University of Campeche, Mexico

Uhl C & Jordan CF (1984) Succession and nutrient dynamics following forest cutting and burning in Amazonia. Ecology 65: 1476–1490

Urabe J, Nakanishi M & Kawabata K (1995) Contribution of metazoan plankton to the cycling of N and P in Lake Biwa. Limnol. Oceanogr. 40: 232–241

Uttormark PD, Chapin JD & Green KM (1974) Estimating nutrient loadings of lakes from non-point sources. United States Environmental Protection Agency Report EPA-660/3-74-020, Washington, DC, U.S.A.

Vallentyne JR (1974) The algal bowl. Canadian Fisheries and Marine Service Miscellaneous Special Publication 22, Ottawa, Canada

Victoria RL, Martinelli, LA, Martatti J & Richey JE (1991) Mechanisms of water recycling in the Amazon Basin: isotopic insights. Ambio. 20: 384–387

Villar CA, de Cabo L & Bonetto CA (1996) Macrophytic primary production and nutrient concentrations In: A Deltaic Floodplain Marsh Of The Lower Paraná River. Hydrobiologia 330: 59–66

Viner AB (1975) The supply of minerals to tropical rivers and lakes (Uganda). In: Hasler AD (Ed.) Coupling of Land and Water Systems (pp 227–261). Springer-Verlag, New York, U.S.A.

Vitousek PM (1984) Litterfall, nutrient cycling, and nutrient limitation in tropical forests. Ecology 65: 1476–1490

Vitousek PM & Howarth RW (1991) Nitrogen limitation on land and in the sea: How can it occur? Biogeochem. 13: 87–115

Vitousek PM & Melillo JM (1979) Nitrate losses from disturbed forests: patterns and mechanisms. Forest Sci. 25: 605–619

Vitousek PM, Aber JD & Tilman DG (1997) Human alteration of the global nitrogen cycle: sources and consequences. Ecological Applications 7: 737–750

Vollenweider RA (1968) Water management research. OECD, Paris, DAS/CSI/68.27

Vorosmarty CJ, Sharma KP, Lough JA (1997) The storage and aging of continental runoff in large reservoir systems of the world. Ambio 26: 210–222

Vought LBM, Dahl J & Lacoursiere JO (1994) Nutrient retention in riparian ecotones. Ambio 23: 342–359

Welsh BL, Whitlatch RB & Bohlen WF (1982) Relationship between physical characteristics and organic carbon sources as a basis for comparing estuarines in southern New England. In: Kennedy VS (Ed.) Estuarine comparisons (pp 53–67). Academic Press, NY, U.S.A.

Williams MR & Melack JM (1997) Solute export from forested and partially deforested catchments in the central Amazon. Biogeochemistry 38: 67–102

Williams MR, Fisher TR & Melack JM (1997a) Chemical composition and deposition of rain in the central Amazon, Brazil. Atmosphere & Environment 31: 207–217

sphere (Vitousek et al. 1997). While of global scope, these perturbations vary regionally (Howarth et al. 1997). In some regions, anthropogenic perturbation of the nitrogen cycle is still small. For present purposes, we define minimally disturbed regions as those with 80% or more natural vegetative cover, population density below 5 indiv. km^{-2} and estimated anthropogenic nitrogen deposition (NO_y-N) below 2.5 kg ha^{-1} y^{-1} (from models: Holland et al. 1997). Flux measurements from such regions, which are becoming increasingly scarce, are the only direct source of information on baseline conditions for the nitrogen cycle over a range of physiographic, hydrologic, and ecological conditions. The purpose of this paper is to summarize and analyze information on the yield of nitrogen compounds in runoff from watersheds in which anthropogenic disturbance has not greatly affected the nitrogen cycle.

The objectives of this paper parallel those of Meybeck (1982), who used a synthesis of information on nitrogen and other elements to reach conclusions about background conditions. The present effort differs from Meybeck's in that it focuses specifically on nitrogen and emphasizes yields (mass $area^{-1}$ $time^{-1}$) rather than concentrations (mass $volume^{-1}$). We deal here only with background conditions, and not with the full range of perturbed conditions, as Meybeck did. Our analysis benefits from recent accumulation of new information on nitrogen yields from minimally disturbed environments. This enriched data base allows for the first time some statistical analysis of factors influencing the natural nitrogen cycle.

The working hypothesis for this analysis is that watersheds, when unperturbed by human influence, will show nitrogen yields that can be predicted on the basis of general environmental variables such as drainage area or amount of runoff, as explained below. The hypothesis is tested empirically with field estimates of nitrogen yield from watersheds in which the nitrogen cycle has been little perturbed by human activities.

Methods

The basis for the present analysis is annual yield of total fixed nitrogen, or of any major components of the total fixed nitrogen pool, from watersheds of known area. Raw data underlying the annual yield estimates include information on concentration and discharge, which together can be used in computing discharge-weighted mean concentrations that can be converted into annual yields. Temporally fragmentary data have been excluded; all of the data considered here span at least one annual cycle and include no fewer than six measurements per cycle for both discharge and concentration. For most sites included in the analysis, information is available on multiple years and includes many more than six measurements per year. Some data sets used in the analysis lack one or more components. Statistical analyses are applied only to data sets for which relevant data are available; thus the number of measurements varies from one statistical test to another.

The analysis includes three independent variables that might be expected to influence nitrogen yield: area, elevation, and annual specific runoff (mm/yr). Area is a surrogate for a cluster of variables, including the time and distance of transport for nitrogen in a drainage network. In this sense, there is a connection between area and the river continuum concept, which postulates changes in organic carbon and nutrients as water passes from one stream order to the next (Allan 1995). Elevation, the second independent variable, is estimated as the mean of the highest and lowest elevations for small watersheds and as the area-weighted mean for large watersheds. Elevation is related to mean gradient and topographic relief, both of which might influence the transit time of nitrogen in soil, and thus the uptake or biologically-induced partitioning of fixed nitrogen. Gradient and relief also are related to erosion, which in turn may influence the transport of particulate nitrogen. Specific runoff (mm/yr), which for convenience is referred to here as runoff, is the third independent variable. It is estimated for years coincident with the water chemistry data used in calculating yields. Specific runoff reflects the flushing rate of the terrestrial system, and by this means may affect nitrogen transport and residence time of fixed nitrogen and nitrogen fractions in the watershed. Because runoff reflects precipitation, soil moisture, and vegetation, it also may be related to a complex of biological factors including potential for biological uptake or processing of nitrogen, and potential for nitrogen fixation.

Few temperate sites and no arctic sites are included here (Table 1). With the exception of studies in the Sierra Nevada of California, data from temperate locations cannot pass the screening criteria because of strong regional enrichment of precipitation in fixed nitrogen, even at relatively long distances from industrial centers. Yield from arctic regions has seldom been measured (but see Meybeck 1982) and presents interpretational difficulties that are difficult to resolve for the present analysis.

Six sites at high elevation in the Sierra Nevada of California are sufficiently free from atmospheric contamination to be part of the data set. The Sierra Nevada data form a distinctive cluster. The reasons for this cannot be resolved because there are too many possibilities, including not only latitude but also elevation (the Sierra Nevada sites are considerably higher than any of the tropical sites included in the data base). Because the Sierra Nevada sites form a distinct population, they have been excluded from the statistical analysis. Therefore, the conclusions of the statistical analysis apply specifically to tropical sites of low to moderate elevation over a wide range of

Table 1. Summary of information on watersheds, sequenced by area, that meet the screening requirements for the data set on nitrogen yields from undisturbed watersheds.

Site	Lat	Long	Elev m asl	Area km^2	Runoff mm y^{-1}	Veg.	Sources[1]
Tropical Watersheds							
Amazon (Óbidos)	5 S	60 W	750	5000000	1200	Forest	i, j
Madeira	8 S	62 W	870	1300000	740	Forest	i, j
Solimões at Içá	3 S	70 W	850	1200000	1200	Forest	i, j
Orinoco	7 N	67 W	800	950000	1300	Mixed	k
Negro	0 S	67 W	200	620000	1600	Forest	i, j
Paraguay River	19 S	57 W	600	363500	123	Savanna	b
Japurá	2 S	68 W	300	300000	2000	Forest	i, j
Juruá	5 S	68 W	200	220000	900	Forest	i, j
Apure	8 N	70 W	550	170000	300	Savanna	k
Guaviare	4 N	69 W	880	150000	1800	Forest	l
Trombetas	0 N	58 W	200	130000	540	Forest	i, j
Meta	6 N	69 W	880	110000	1600	Savanna	l
Caroní	6 N	63 W	800	95000	1700	Forest	k
Caura	6 N	66 W	800	47500	2400	Forest	m
Gambia River	13 N	14 W	500	42000	110	Savanna	c
Icacos	18 N	66 W	700	3.20	3683	Forest	h
Rio Tempisquito	10 N	85 W	960	3.19	2900	Forest	g
Rio Tempisquito Sur	10 N	85 W	915	3.11	4300	Forest	g
Quebrada Kathia	10 N	85 W	1070	2.64	3100	Forest	g
Sonadora	18 N	66 W	700	2.60	2497	Forest	h
Quebrada El Jobo	10 N	85 W	815	0.55	1400	Forest	g
Quebrada Zompopa	10 N	85 W	850	0.37	2800	Forest	g
Quebrada Marilin	10 N	85 W	735	0.36	1600	Forest	g
Braço do Mota	3 S	61 W	100	0.18	1650	Forest	e
Tóronja	18 N	66 W	400	0.16	1750	Forest	h
Temperate Watersheds							
Emerald Lake	36 N	189 W	2800	1.20	854	Alpine	a
Spuller Lake	38 N	119 W	3130	1.00	789	Alpine	a
Log Creek	37 N	119 W	2070	0.50	219	Forest	d
Tharp's Creek	37 N	119 W	2070	0.13	219	Forest	d
Sierra Lake 1	37 N	119 W	2960	0.01	352	Alpine	f
Sierra Lake 2	37 N	119 W	2960	0.00	595	Alpine/ rock	f

[1] Sources: a: Melack et al. 1997; b: Hamilton et al. 1997; c: Lesack et al. 1984; d: Williams and Melack 1997a, b; e: Lesack 1993, Williams et al. 1997; f: Williams 1997; g: Newbold et al. 1995; h: McDowell and Asbury 1994; i: Richey unpublished; j: Lewis et al. 1995; k: Lewis and Saunders 1990; l: Weibezahn 1990; m: Lewis et al. 1986

watershed sizes, moisture regimes, and vegetation types. The Sierra Nevada data are tabulated as a means of providing perspective on the data for tropical locations.

Results

Table 2 summarizes the annual yields of nitrogen and nitrogen fractions for watersheds in this analysis. Fewer than half of the studies that meet the criteria for the analysis include all major fractions of total fixed nitrogen. Measurements of nitrate are most common, whereas measurements of ammonium, dissolved organic nitrogen, and particulate nitrogen are less common. Estimates of total nitrogen yield are available for 17 of the 31 watersheds shown in Table 2.

The three independent variables for this analysis (elevation, area, runoff) were tested for statistical relationships to each other (log-log transformation followed by regression analysis). Elevation is unrelated to runoff or to area ($p > 0.05$); area is significantly but weakly ($r^2 = 0.26$) related to runoff: log (runoff, mm y^{-1}) = 3.34 – 0.071 log (area, km^2). In this data set, large watersheds tend to have less runoff per unit area than small watersheds.

A multiple regression analysis following log-log transformation of all variables shows that, for total nitrogen and each of the nitrogen fractions shown in Table 2: (1) significant variance can be explained ($p < 0.05$) by the three independent variables, (2) runoff is consistently the dominant variable explaining variance, and (3) when runoff is used first in multiple regression analysis, area explains no significant added variance and elevation explains significant added variance only for dissolved inorganic nitrogen and its components. For these reasons, the remainder of the analysis can be simplified by focus on relationships between yield and runoff, except in the case of dissolved inorganic nitrogen and its components, for which elevation is also important (Table 3).

Inorganic nitrogen

Inorganic nitrogen consists primarily of ammonium and nitrate. Although the ammonium ion will be in equilibrium with ammonium hydroxide and ammonia, ammonium predominates at the pH of most surface waters (Erickson 1985). For present purposes, the composite of all forms will be designated as ammonium. In principle, nitrite and gaseous nitrogen oxides also could be present. Broad experience with the analysis of nitrite in surface waters has shown that nitrite is a negligible contributor to inorganic nitrogen in flowing waters in the absence of wastewater or anaerobic water sources

Table 2. Nitrogen yields for the watersheds listed in Table 1 (sources as in Table 1). Totals for DIN, TDN, and TN may not equal the sum of fractions if totals were analyzed separately from fractions.

Site	kg ha^{-1} y^{-1}						
	NH_4-N	NO_3-N	DIN	DON	TDN	PN	TN
Tropical Watersheds							
Amazon (Óbidos)	0.24	1.68	1.92	1.94	3.86	2.42	6.06
Madeira		1.30	1.30	1.30	2.60	2.03	4.63
Solimões at Içá		2.40	2.40	1.36	3.76	3.90	7.66
Orinoco	0.33	1.04	1.50	2.08	3.58	2.41	5.98
Negro		0.67	0.67	2.48	3.15	0.02	1.47
Paraguay River					0.57	0.16	0.73
Japurá		2.04	2.04	1.84	3.88	2.44	8.60
Juruá		1.82	1.82	1.21	3.02	1.44	4.45
Apure	0.13	0.41	0.53	0.80	1.33	1.54	2.87
Guaviare	0.31	0.94	1.24				
Trombetas			0.53				3.01
Meta	0.37	1.78	2.14				
Caroní	0.61	1.46	2.07	2.35	4.42	0.75	5.17
Caura	0.89	1.49	2.38	4.00	6.38	3.64	9.98
Gambia River					1.11	0.15	1.26
Icacos	0.68	2.54	3.22	4.79	8.01	1.80	9.80
Rio Tempisquito		6.10	6.10	3.00	9.10		
Rio Tempisquito Sur		4.90	4.90	2.10	7.00		
Quebrada Kathia		5.60	5.60	1.50	7.10		
Sonadora	0.30	1.39	1.69	3.74	5.43	0.47	5.90
Quebrada El Jobo		4.30	4.30	2.00	6.30		
Quebrada Zompopa		6.00	6.00	3.40	9.40		
Quebrada Marilin		4.00	4.00	1.90	5.90		
Braço do Mota	0.14	0.60	0.74	2.87	3.61	0.70	4.31
Tóronja	0.26	0.90	1.16	2.80	3.96	0.48	4.40
Temperate Watersheds*							
Emerald Lake	0.04	0.56	0.60				
Spuller Lake	0.01	0.45	0.45				
Log Creek	0.01	0.00	0.02				
Tharp's Creek	0.01	0.01	0.01				
Sierra Lake 1	0.01	0.03	0.04				
Sierra Lake 2	0.04	0.07	0.11				
Mean (tropical only)	0.39	2.43	2.53	2.37	4.70	1.52	5.08
Number of sites	11	22	23	20	22	16	17
Standard error	0.23	1.80	1.73	1.00	2.40	1.18	2.71

* Excluded from statistical summaries.

Table 3. Summary of the relationship between total nitrogen and various nitrogen fractions and amount of runoff (log-log). All relationships are statistically significant at $p < 0.05$. The table also shows, based on the statistical relationships, the expected yields and volume-weighted mean concentrations of the nitrogen fractions at low runoff (e.g., Gambia), at the savanna-forest transition (400 mm), in the mid-range of forest runoff (2000 mm), and at highest runoff (4000 mm).

Fraction or Ratio	Const	Slope	Std. err. slope	r^2	N (kg ha^{-1} y^{-1}) at runoff (mm y^{-1})				N ($\mu g\ l^{-1}$) at runoff (mm y^{-1})			
					100	400	2000	4000	100	400	2000	4000
Ammonium-N*	−2.57	0.65	0.23	0.46	0.05	0.13	0.38	0.59	54	33	19	15
Nitrate-N*	−2.33	0.80	0.24	0.36	0.19	0.56	2.05	3.56	186	141	102	89
Dissolved inorg. N*	−2.41	0.84	0.19	0.49	0.19	0.60	2.31	4.13	186	149	115	103
Dissolved org. N	−1.44	0.55	0.11	0.57	0.46	0.98	2.37	3.48	457	245	119	87
Total diss. N	−1.50	0.67	0.05	0.89	0.69	1.75	5.15	8.19	692	438	257	205
Particulate N**	−1.68	0.58	0.22	0.34	0.30	0.67	1.72	2.57	302	169	86	64
Total N**	−1.24	0.63	0.07	0.85	1.05	2.51	6.91	9.45	1047	627	376	267

* Significant added variance for elevation (see text).
** Excludes R. Negro.

(Meybeck 1982; Stumm & Morgan 1996). The reactive oxides of nitrogen (primarily NO and NO_2) cannot long persist in water. N_2O, which is much less reactive, is of interest as a greenhouse gas but occurs at very low concentrations and thus does not contribute much to the mass transport of fixed nitrogen. Therefore, we treat the sum of ammonium and nitrate as dissolved inorganic nitrogen (DIN).

Table 2 shows the mean yield of ammonium (0.39 kg ha^{-1} y^{-1}; 23% of DIN) at the 11 sites for which information on ammonium is available. Some studies excluded ammonium after initial analyses showed it to be present only at very low concentrations. In this sense, the mean in Table 2 may be biased toward locations that have easily measurable concentrations of ammonium. If missing values for ammonium are treated as zeros, ammonium accounts for about 15% of total inorganic nitrogen. Thus the true mean is probably between 15% and 23%; we assign it a nominal value of 20%.

Yield of ammonium increases with yield of nitrate (log NH_4-N kg ha^{-1} y^{-1} = 0.91 log NO_3-N kg ha^{-1} y^{-1} −0.54; r^2 = 0.62). The coefficient of the log-log relationship is not significantly different from 1.0, i.e., the ratio of ammonium to DIN is essentially constant over a broad range of DIN concentrations. Yield of ammonium is positively related to amount of runoff; it increases about 50% when runoff doubles (Table 3). Secondarily, it is also positively related to elevation (R^2 = 0.73 for runoff plus elevation; for runoff alone, r^2 = 0.46).

As shown by Table 2, the yield of nitrate averages 2.4 kg ha^{-1} y^{-1}, and is strongly and positively related to runoff (Table 3, Figure 1). When both runoff and elevation are included as independent variables in a multiple regression analysis (log-log), R^2 = 0.49, while runoff alone gives r^2 = 0.36.

Because the trends in yield of ammonium and nitrate are similar, both are reflected in yields of DIN, which shows a strong positive relationship with amount of runoff, and a weaker relationship with elevation (R^2 = 0.65 for runoff and elevation together, r^2 = 0.55 for runoff alone).

Yields of dissolved inorganic nitrogen passing from the Sierra Nevada watersheds, which are excluded from the statistical analysis of Table 3, are extraordinarily low (all $\leq$ 0.6 kg ha^{-1} y^{-1}; Table 2). These watersheds are uniformly small, but they overlap in size with small tropical watersheds that have much higher yields. Although the runoff for these watersheds is among the lowest for the data set, the inorganic nitrogen yields are far lower than would be expected from the rest of the data set on the basis of runoff (Figure 1).

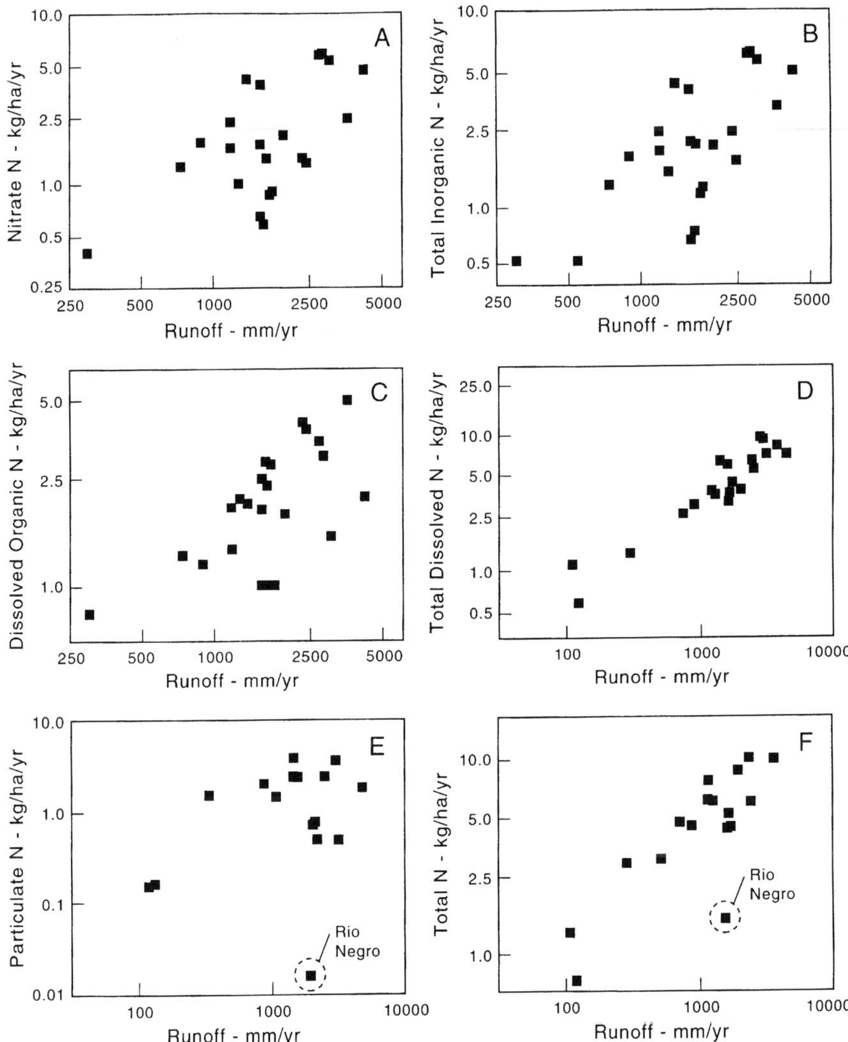

Figure 1. Relationship of annual runoff to annual yields of nitrogen fractions and total nitrogen in minimally disturbed tropical watersheds.

Dissolved organic and total dissolved nitrogen

The mean yield of dissolved organic nitrogen (DON) from the 20 watersheds in which it was measured is 2.4 kg ha^{-1} yr^{-1}, i.e., the same as the yield of DIN. Yield of DON is positively related to amount of runoff (Table 3). Yield of total dissolved nitrogen (TDN) also shows a strong positive relationship to runoff (Table 3, Figure 1). The exponent for the log-log relationship is significantly ($p < 0.05$) below 1: yield of dissolved nitrogen increases about two-thirds as fast as runoff.

Particulate nitrogen

Yield of particulate nitrogen (PN) averages 1.5 kg ha^{-1} y^{-1} across the 16 watersheds for which it was measured. A plot of the relationship between yield of particulate nitrogen and runoff shows that the Rio Negro has an anomalously low yield of particulate nitrogen (Figure 1), which reflects its extraordinarily low load of suspended solids. If the Rio Negro is removed as a special case, there is a significant positive relationship between particulate nitrogen and amount of runoff (Table 3).

Total nitrogen

Yield of total nitrogen (TN) is closely and positively related to the amount of runoff (Figure 1), but increases at only about two-thirds the rate of runoff (Table 2). The Rio Negro is anomalous because of its exceptionally low yield of particulate nitrogen, as mentioned above.

Ratios

The ratios DON/TDN and PN/TN were analyzed statistically. Conclusions about complementary ratios (DIN/TDN and TDN/TN) follow directly from the two ratios that were analyzed.

The ratios DON/TDN and PN/TN are, as shown by multiple regression analysis (ratios untransformed, independent variables log transformed), significantly related to area but not to elevation or runoff after area is taken into account. The mean of the ratio DON/TDN is 0.50 (std deviation, 0.17); larger watersheds have proportionately less DON in the TDN fraction (Figure 2, Table 4). The mean of the ratio PN/TN is 0.30 (standard deviation, 0.16); larger watersheds have proportionately more PN (Table 4).

Discussion

Yields of total nitrogen and all nitrogen fractions show a strong and positive relationship to amount of runoff. Yields increase less rapidly than runoff, however, and this accounts for a decline in discharge-weighted mean concentrations of total nitrogen and nitrogen fractions with increasing annual runoff (Table 3). Because runoff is correlated with vegetation, the relationship between nitrogen yield and runoff corresponds to a relationship between

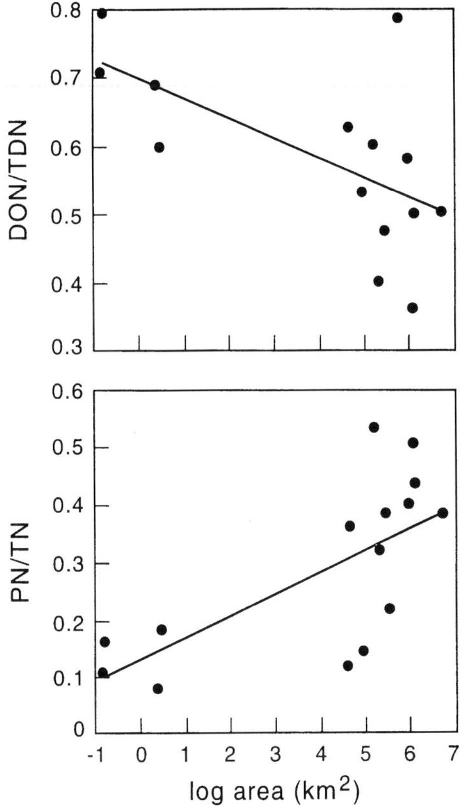

Figure 2. Relationship of watershed area to ratios DON/TDN and PN/TN for minimally disturbed tropical watersheds.

Table 4. Relationships between watershed area and the ratios of nitrogen fractions (area is log transformed, ratio is not transformed).

					Ratios by Stream Order**					
					2	4	6	8	10	12
					Ratios by Area (km^2)					
Ratio	Constant	Slope	SE Slope	r^2	10	10^2	10^3	10^4	10^5	10^6
DON/TDN	0.70	−0.029	0.011	0.38	0.67	0.64	0.61	0.58	0.55	0.52
PN/TN*	0.13	0.034	0.014	0.49	0.17	0.21	0.25	0.29	0.33	0.37

* R. Negro omitted.
** As determined approximately from area.

mangrove growth (Feller 1995). Both studies showed that mangroves in these sites are also influenced by a combination of fertility and hydroperiod. Fertilization studies of scrub mangroves in south Florida have also shown that phosphorus rather than nitrogen can stimulate growth in red mangroves (Koch 1996). A Monod model (Bridgham et al. 1995) of forest biomass along the estuarine gradient of the Shark River estuary has a stronger correlation with soil concentrations of P than N (Chen 1996), with a half-saturation constant of 29.3 g/m^2 (to a depth of 40 cm).

The lack of nitrogen limitation may be associated with the accumulation of this nutrient in leaf litter supplied either from the atmosphere and/or from tidal waters. It is a general pattern to observe a net increase in nitrogen during the initial two months of decomposing leaf litter that reduces the C:N ratio of this material to nearly half the ratios of litter falling from the canopy (Twilley et al. 1986; Twilley et al. 1997). Nitrogenase activity has been observed in decomposing leaves, root surfaces (prop roots and pneumatophores) and sediment, but few studies have interpreted areal fixation rates that can be compared to nitrogen demand by annual net production of mangrove biomass (Kimball & Teas 1975; Gotto & Taylor 1976; Zuberer & Silver 1978; Potts 1979; Gotto et al. 1981). Results from mangrove sediments in south Florida indicate that nitrogen fixation rates range from 0.4 to 3.2 $g\ N.m^{-2}.y^{-1}$ (Kimball & Teas 1975; Zuberer & Silver 1978), similar to the natural rates of denitrification. These studies have shown that decaying mangrove leaves are sites of particularly high rates of fixation, and thus may account for some of the nitrogen immobilization in leaf litter on the forest floor (Gotto et al. 1981; Van der Valk & Attiwill 1984). In addition, there seems to be a pulse of nitrogen fixation in mangrove leaf litter with a 20 day lag following the initiation of decomposition on the forest floor (Pelegri et al. 1997). This may result from a combination of factors including organic matter source, time for microbial colonization, and inhibition by phenolics (Pelegri and Twilley 1998). However, the spatial and temporal analysis of nitrogen fixation is still inadequate to provide a clear estimate of this contribution to the nitrogen budget of mangrove wetlands.

While a substantial capacity has been documented for denitrification of nitrate-laden secondary effluents in mangrove sediments (Corredor & Morell 1994), doubt exists as to the capacity of microbial communities in these sediments to oxidize reduced nitrogen. Rivera-Monroy and Twilley (1996), documented only minimal losses of ^{15}N-labeled reduced organic substrates when added to mangrove sediments of Laguna de Terminos on the Gulf of Mexico. These data thus indicate minimal transformation to gaseous products of the added substrate and consequently of a lack of coupled nitrification/denitrification. Competitive advantage in reduced nitrogen uptake of heterotrophic bacterial communities over the nitrifiers, or the inhibition of nitrifying activity by phenolic compounds or reduced sulfur may in part explain these results. In mangrove sediments devoid of sulfide, nitrification can proceed at rates of *ca.* 250 μmole $N.m^{-2}.h^{-1}$ (Corredor et al. In Press) among the highest values recorded for this process throughout the marine environment (Kaplan 1983). It must be held in mind however that in these particular environments, the mangroves themselves can compete vigorously for both the reduced and oxidized species of available nitrogen. The lack of $^{15}N_2$ gas production indicates that much of the net DIN exchange at the boundary of mangroves (Rivera-Monroy et al. 1995; Twilley 1997) may not be lost to the atmosphere via denitrification, but accumulated in the litter on the forest floor due to the high demand associated with C:N ratios that can exceed 70 in senescent leaf litter (Twilley et al. 1986).

The relative availability of nitrogen and phosphorus to mangrove production is recorded in the concentration and burial of these two nutrients in mangrove sediments. Mangroves in river-dominated estuaries have higher sedimentation rates of inorganic matter with N:P ratios <10 compared to mangroves in reef environments with little terrigenous input and N:P ratios >60 (Twilley 1995, 1997). The availability of P associated with the deposition of terrigenous sediments can be linked to the litter productivity of mangrove wetlands. Nitrogen concentrations vary directly with organic content of mangrove sediments and exhibit less variation among sites than concentrations of P. Thus in nearshore oligotrophic waters of the Caribbean, there seems to be less phosphorus that nitrogen due to the relative availability of terrigenous *vs* atmospheric sources for these two nutrients. A more thorough examination of these unique ecosystems is certainly due.

Anthropogenic impact

Human activity is causing large imbalances in the nitrogen and phosphorus cycles of coastal marine waters as a consequence of increased fertilizer use, of wetland reclamation for aquaculture ponds, of poor land management practices leading to topsoil erosion, and of large inputs of treated and untreated sewage. Abnormal nitrogen and phosphorus inputs to otherwise nutrient limited coastal waters can bring about rapid changes in algal communities disrupting existing ecological structure. Opportunistic algal species, particularly the filamentous cyanophytes and several members of the Chlorophyta may come to dominate the benthos excluding both other macrophytes and the macroinfauna as well. In Puerto Rico, Diaz et al. (1990) and Corredor et al. (1992) have documented the course of environmental degradation brought about by the filamentous cyanophyte *M. lyngbyaceus* in response to anthropogenic eutrophication. Monospecific mats of this species cover large areas of the shallow benthos in affected areas displacing the previously predominant marine grasses (mainly *Thalassia testudinum*). *M. lyngbyaceus* mats are buoyed during the daytime by oxygen entrapped within the dense mat and sink to the bottom during the night when the entrapped oxygen is consumed by the microbial and faunal populations of the mat community. When the mats sinks to a depth below the euphotic zone, already narrowed by the proliferation of planktonic algae, decomposition of the mats brings about severe anoxia of the sediments with the consequent exclusion the natural infaunal communities.

In coral reef environments, fleshy green algae and can rapidly overwhelm the coral population in response to excessive anthropogenic nutrient inputs as was demonstrated for Hawaiian reefs by Smith (1984). In coral reefs of the tropical Intra-American seas, proliferation of fleshy and filamentous macroalgae have also been reported. Littler et al. (1992) found significant growth of *Sargassum sp* and *Enteromorpha sp.* associated to anthropogenic nutrient inputs off Fort-de France, Martinique while Lapointe et al. (1992) reported extensive populations of *Ulva, Chaetomorpha* and others along the Belize barrier reef associated with natural eutrophication from bird rookeries. While in both cases, alleviation of phosphorus limitation was implicated, increased nitrogen inputs were reported and concurrent alleviation of nitrogen limitation may not be ruled out. Healthy coral reef ecosystems, set in clear oligotrophic waters experience low sedimentation of detrital organic material and diffusion of remineralized nutrients from the sediments is moderate (Corredor & Morell 1985). Nevertheless, sessile organisms of these ecosystems filter or capture large volumes of plankton representing a large input of reduced nitrogen. While ammonium release from the reef framework is low, Corredor et al. (1988) have shown that sponges, the principal filtering organisms on the reef, excrete nitrate rather than ammonium; presumably as a result of the activity of symbiotic populations of nitrifying bacteria within the sponges. As planktonic and macrobenthic algae exhibit a marked preference for reduced nitrogenous species over the oxidized forms this mechanism might serve to maintain algal populations in check. When water quality deteriorates, however, and increased inputs of dissolved and detrital nutrients appear, coral reefs rapidly succumb to the competition of the fleshy algae.

Nitrification and denitrification are other benthic process which may function differently in nearshore subtidal tropical marine ecosystems than in their temperate counterparts. In temperate estuaries, denitrification plays a major role in reducing the effects of increased nitrogen loading from human activity, with much of the denitrification resulting from close coupling to nitrification (Seitzinger 1988; Nixon et al. 1996). In contrast, microbial oxidation of ammonium (nitrification) in tropical coastal sediments appears to be more severely constrained than in comparative temperate environments (Morell & Corredor 1993). Oxygen requirements of nitrifying bacteria (whose activity peaks in microaerophyllic, environments) and strong inhibition by light may explain the fragility of these communities in tropical marine environments. Susceptibility to light limits nitrification to the lower sediment layers below the limits of light penetration (Vanzella et al. 1989); light is less of a problem in most temperate estuaries because of the greater turbidity. Oxygen depletion of subsurface sediments poses a lower limit for active nitrification in tropical sediments. Hydrogen sulfide can aggravate this lower limit, both by consuming oxygen and by inhibiting the enzyme responsible nitrification (Richardson 1985). Given comparable levels of respiration, hydrogen sulfide concentrations will build up more quickly in tropical carbonate sediments than in temperate estuarine sediments because of the much lower concentrations of iron available for precipitating the sulfides. Thus, in comparison to temperate estuarine sediments, the realm available to nitrifying bacteria in many tropical lagoons is reduced to a thin and vulnerable veneer where light does not penetrate and yet where oxygen is present and sulfides are absent (Morell & Corredor 1993). In coral reef sands, nitrification and denitrification proceed at comparable rates (Corredor & Capone 1985), thus maintaining low levels of ammonium in the sediments. However, eutrophication will increase oxygen consumption in the sediment, increase the production of sulfides through sulfate reduction, and thereby decrease the zone in the sediment where nitrification can occur (Morell & Corredor 1993). Given sufficient input of organic matter to the sediment, nitrification and denitrification both cease. As a result, interstitial ammonium concentrations may increase to very high levels (millimolar range), and the ammonium flux from sediment to overlying water column can be high (Mosquera et al. 1998).

The carbonate sands (calcite and aragonite) which often dominate tropical lagoons provide a major sink for phosphorus. Phosphate readily adsorbs onto these sediments, which contributes to phosphorus limitation in many tropical seagrass systems (Morse et al. 1985, 1987; Short et al. 1990; McGlathery et al. 1994; Howarth et al. 1995). Although the affinity of phosphate for carbonate minerals is in fact less than that for iron minerals found in the surface layers of temperate estuarine sediments (Krom & Berner 1980), the huge mass of carbonates in tropical lagoons provides a substantial sink for phosphate. However, the rate of phosphate uptake by carbonate sands appears to be slower in more eutrophic sites than in oligotrophic sites. Thus, as more phosphorus is added into a tropical lagoon, the phosphorus becomes proportionally more available as the rate of sorption onto the sediment slows (McGlathery et al. 1994; Howarth et al. 1995). This can result in a conversion from phosphorus to nitrogen limitation in the ecosystem (McGlathery et al. 1992, 1994; Howarth et al. 1995). The mechanism for the slower uptake of phosphate in eutrophied tropical systems is still not entirely understood (McGlathery et al. 1994). Most of the phosphorus in carbonate sands is present as carbonate-fluoro-apatite (Jensen et al. 1997). Apparently, once phosphate adsorbs onto the surface of the sands, it is slowly transformed into the apatite mineral, and this conversion may replenish the active sites for phosphate adsorption. If so, during eutrophication as the apatite content of the sediment increases, the rate of new apatite formation and therefore the rate of replenishment of surface-sorption sites for phosphate may slow (McGlathery et al. 1994; Jensen et al. 1997).

In summary, nitrogen and phosphorus both exert substantial influence on biological productivity and structure of tropical marine ecosystems. As offshore waters are nitrogen limited, the transition to phosphorus limitation in nearshore environments would appear to be a function of the residence time of water masses and their consequent exposure to sediments capable of immobilizing phosphorus. The nature of the sediment, the mixing rate between nutrient-laden runoff waters and nutrient-poor oceanic waters and the degree of interaction of these water masses with the sediment will probably control the dynamics of this transition.

Nearshore tropical marine ecosystems function differently from their temperate counterparts where coupled nitrification/denitrification serves as an important mechanism for nitrogen depuration. In contrast, nearshore tropical ecosystems are more susceptible to nitrogen loading as depurative capacity of the microbial communities is limited by the fragility of the nitrification link. At the same time, accumulation of organic matter in nearshore carbonate sediments appears to impair their capacity for phosphorus immobilization. In the absence of depurative mechanisms for either phosphorus or nitrogen, limitation for both these nutrients is alleviated and continued nutrient loading fuels the proliferation of nuisance algae.

Disruption of the marine nitrogen cycle can bring about not only deterioration of the aquatic environment but extends to the atmosphere when these imbalances favor the emission of gaseous products. Nitrous oxide, an intermediate of both nitrification and denitrification, is a potent green house gas and is, moreover, active in the destruction of stratospheric ozone (Delwiche 1981). Marine environments are normally a source rather than a sink for nitrous oxide but tropical waters, with the notable exception of upwelling ecosystems (Law & Owens 1980; Naqvi & Noronha 1991), exhibit the lowest rates of emission (Weiss 1981; Nevisson et al. 1995); in large part due to the depletion of available nitrogen in surface waters and the barrier to diffusion from the supersaturated lower layers posed by the well stratified surface layer.

Coastal and intertidal waters, particularly those subject to eutrophication, can however emit nitrous oxide to the atmosphere at rates significantly above those of the open ocean. Relatively pristine mangrove sediments normally emit nitrous oxide at rates up to 10 times those observed offshore. These emissions may in turn increase dramatically as allochthonous nitrogen is provided. Corredor et al. (In Press) have demonstrated enhanced nitrous oxide emissions from mangrove sediments subject to well nitrified sewage releases and greater rates yet in mangrove sediments receiving reduced nitrogenous inputs. Large scale modification of mangrove ecosystems in the neotropics thus has the potential for contributing further to global climate change.

Evidently, our knowledge of nitrogen cycling in tropical marine ecosystems is but a loosely interconnected mosaic. A sustained and concerted effort is required if we are to acquire a well rounded understanding of these important processes.

Acknowledgements

This paper was initially prepared for presentation at the SCOPE Nitrogen workshop in Termas de Chillán, Chile. J.E. Corredor and J.M. Morell gratefully acknowledge support from NASA through Grants No. NAGW 3926 and NCCW-56. R.W. Howarth was partially supported by NSF grant DEB 9527405. R.R. Twilley acknowledges support from the U.S. Environmental Protection Agency (Grant No. CR820667) and the U.S. Department of Interior (Grant No. CS5280-4-9019).

References

Aller RC (1988) Benthic fauna & biogeochemical processes in marine sediments: the role of burrow structures. In: Blackburn TH & Sorensen J (Eds) Nitrogen Cycling in Coastal Marine Eenvironments. Wiley, NY, U.S.A.

Belser LW & Mays EAL (1980) Specific inhibition of nitrite oxidation by chlorate and its use in assessing nitrification in soils and sediments. Appl. Environ. Microbiol. 39: 505–510

Bergman B, Gallon JR, Rai AN & Stal LJ (1997) N_2 fixation by non-heterocystous cyanobacteria. FEMS Microbiology Reviews 19: 139–185

Bonilla JW, Senior J, Bugden O, Zafiriou & Jones R (1993) Seasonal distribution of nutrients and primary productivity on the eastern continental shelf of Venezuela as influenced by the Orinoco River. J. Geophys. Res. 98: 2245–2257

Bushaw KL, Zepp RG, Tarr MA, Schultz-Janders D, Bourbonniere RA, Hodson RE, Miller WL, Bronk DA & Moran MA (1996) Photochemical release of biologically available nitrogen from aquatic dissolved organic matter. Nature 381: 404–407

Capone DG, Zehr JP, Paerl HW, Bergman B & Carpenter EJ (1997) *Trichodesmium*, a globally significant marine cyanobacterium. Science 276: 1221–1229

Capone DG (1983) Benthic nitrogen fixation. In: Carpenter EJ & Capone DG (Eds) Nitrogen in the Marine Environment (pp 105–137). Academic

Carlsson P, Segatto AZ & Granéli E (1993) Nitrogen bound to humic matter of terrestrial origin – a nitrogen pool for coastal phytoplankton? Mar. Ecol. Prog. Ser. 97: 105–116

Carpenter EJ & Capone DG (1992) Nitrogen fixation in *Trichodesmium* blooms. In: Carpenter EJ, Capone DG & Rueter JG (Eds) Marine Pelagic Cyanobacteria: *Trichodesmium* and Other Diazotrophs (pp 211–217). Kluwer Academic Publishers, Dordrecht, The Netherlands

Carpenter EJ (1983) Nitrogen fixation by marine *Oscillatoria (Trichodesmium)* in the world's ocean. In: Carpenter EJ & Capone DG (Eds) Nitrogen in the Marine Environment (pp 65–104). Academic

Chen R (1996) Ecological analysis and simulation models of landscape patterns in mangrove forest development and soil characteristics along the Shark River estuary, Florida. PhD Dissertation, University of Southwestern Louisiana, Lafayette, Louisiana, U.S.A.

Clark LL, Ingall ED & Benner R (1998) Marine phosphorus is selectively remineralized. Nature 393: 426

Corredor JE, Morell JM & Díaz MR (1994) Environmental degradation, nitrogen dynamics and proliferation of the filamentous cyanophyte *Microcoleus lyngbyaceus* in nearshore Caribbean waters. In: Ezio Amato (Ed.) Mediterraneo e Caraibe due mari in pericolo? Sversamenti accidentali di idrocarburi ed emergenze causate dalle alghe. ICRAM/IFREMER. Atti del Convegno Internazionale. Genova 1992

Corredor JE (1976) Aspects of phytoplankton dynamics in the Caribbean Sea. FAO Fisheries Report No. 200: 101–114

Corredor JE (1979) Phytoplankton response to low-level nutrient enrichment through upwelling in the Colombian Caribbean Basin. Deep-Sea Res. 26A: 731–741

Corredor JE & Capone DG (1985) Studies on nitrogen diagenesis in coral reef sands. In: Gabrie C et al. (Eds) Proc. Vth Intnl. Coral Reef Symp. 3: 395–399. Papeete, Tahiti, French Polynesia

Corredor JE & Morell J (1985) Inorganic nitrogen in coral reef sediments. Mar. Chem. 16: 379–384

Corredor JE & Morell JM (1994) Nitrate depuration of secondary sewage effluents in mangrove sediments. Estuaries. 17(2): 295–300

Corredor JE & Morell JM (1989) Assessment of inorganic nitrogen flux across the sediment-water interface in a tropical lagoon. Est Coast Shelf Sci 28: 339–345

Corredor JE, Morell J & Mendez A (1984) Dissolved nitrogen, phytoplankton biomass and island mass effects in the northeastern Caribbean Sea. Carib. J. Sci. 20: 129–137

Corredor JE, Wilkinson CR, Vicente VP, Morell JM & Otero E (1988) Nitrate release by Caribbean reef sponges. Limnol. Oceanogr. 33: 114–120

Corredor JE, Morell JM & Bauzá J (In Press) Atmospheric nitrous oxide flux from mangrove sediments. Mar. Poll. Bull.

Corredor JE & Morell JM (In Review) Seasonal variation of physical and biogeochemical features is Eastern Caribbean surface water. Submitted to J. Geophys. Res.

Delwiche CC (1981) The nitrogen Cycle & Nitrous Oxide. In: Delwiche CC (Ed.) Denitrification, Nitrification and Atmospheric Nitrous Oxide (pp 1–84). John Wiley & Sons, NY, U.S.A.

Diaz MR, Corredor JE & Morell JM (1990) Inorganic nitrogen uptake by *Microcoleus lynbyaceus* mat communities in a semi-eutrophic marine community. Limnol. Oceanogr. 35: 1788–1795

Downing JA (1997) Marine nitrogen:phosphorus stoichiometry and the global N:P cycle. Biogeochemistry 37: 237–252

Feller IC (1995) Effects of nutrient enrichment on growth and herbivory of dwarf red mangrove (*Rhizophora mangle*). Ecological Monographs 65: 477–505

Garret C & Munk W (1972) Oceanic mixing by breaking internal waves. Deep-Sea Res. 19: 823–832

Gotto JW & Taylor BF (1976) N_2 fixation associated with decaying leaves of the red mangrove (*Rhizophora mangle*). Applied and Environmental Microbiology 31: 781–783

Gotto JW, Tabita FR & Baalen CV (1981) Nitrogen fixation in intertidal environments of the Texas gulf coast. Estuarine, Coastal and Shelf Science 12: 231–235

Howarth RW & Marino R (1998) A mechanistic approach to understanding why so many estuaries and brackish waters are nitrogen limited. In: Hellstrom T (Ed.) Effects of Nitrogen in the Aquatic Environment. Swedish National Committee for IAWQ, Royal Swedish Academy of Sciences, Stockholm, in press

Howarth R, Billen WG, Swaney D, Townsend A, Jaworski N, Lajtha K, Downing JA, Elmgren R, Caraco N, Jordan T, Berendse F, Freney J, Kudeyarov V, Murdoch P & Zhao-Liang Zhu (1996) Regional nitrogen budgets and riverine N & P fluxes for the drainages to the North Atlantic Ocean: Natural and human influences. Biogeochemistry 35: 75–139

Howarth RWHS, Marino Jensen R & Postma H (1995) Transport to and processing of P in near-shore and oceanic waters. In: Tiessen H (Ed) Phosphorus in the Global Environment (pp 323–345). Wiley & Sons, Chichester

Howarth RW, Marino R & Chan F. What regulates nitrogen fixation by planktonic cyanobacteria in aquatic ecosystems? This volume

Howarth RW, Marino R, Lane J & Cole JJ (1988) Nitrogen fixation in freshwater, estuarine, and marine ecosystems. 1. Rates and importance. Limnol. Oceanogr. 33: 669–687

Howarth RW (1988) Nutrient limitation of net primary production in marine ecosystems. A. Rev. Ecol. 19: 89–110

Jensen H, McGlathery KJ, Marino R & Howarth RW (1998) Forms and availability of sediment phosphorus in carbonate sand of Bermuda seagrass beds. Limnol. Oceanogr. 43: 799–810

Kaplan WA (1983) Nitrification. In: Carpenter EJ & Capone DG (Eds) Nitrogen in the Marine Environment (pp 139–190). Academic, NY, U.S.A.

Kimball MC & Teas HJ (1975) Nitrogen fixation in mangrove areas of southern Florida. In: Walsh G, Snedaker S & Teas H (Eds) Proceedings of the International Symposium on the Biology and Management of Mangroves (pp 654–660). Institute of Food and Agricultural Sciences University of Florida, Gainesville, Florida, U.S.A.

Koch MS (1996) Resource availability and abiotic stress effects on *Rhizophora Mangle L.* (Red Mangrove) development in South Florida. PhD Dissertation, University of Miami, Coral Gables, Florida, U.S.A.

Krom MD & Berner RA (1980) Adsorption of phosphate in anoxic marine sediments. Limnol. Oceanogr. 25: 797–806

Lapointe B, Littler MM & Littler DS (1992) Modification of benthic community structure by natural eutrophication: The Belize Barrier reef. In: Richmond RH (Ed.) Proc. 7th Coral Reef Symp. Guam. 1: 323–334

Law CS & Owens JP (1990) Significant flux of atmospheric nitrous oxide from the northwest Indian Ocean. Nature 346: 826–828

Lewis WM & Saunders JF III (1989) Concentration and transport of dissolved and suspended substances in the Orinoco River. Biogeochemistry 7: 103–240

Littler MM, Littler DS & Lapointe B (1992) Modification of tropical reef community structure due to cultural eutrophication: The southwest coast of Martinique. In: Richmond RH (Ed.) Proc. 7th Coral Reef Symp. Guam. 1: 335–343

Lynch JC, Meriwether JR, McKee BA, Vera-Herrera F & Twilley RR (1989) Recent accretion in mangrove ecosystems based on ^{137}Cs and ^{210}Pb. Estuaries 12: 284–299

Margalef R (1965) Composición y distribución del fitoplancton In: Estudios sobre el Ecosistema Pelágico del N.E. de Venezuela. Mem. Fund. Cienc. Nat. La Salle 25: 141–208

Margalef R (1971) The pelagic ecosystem of the Caribbean Sea. In: Symposium on Investigations and Resources of the Caribbean Sea and Adjacent Regions (pp 483–489). UNESCO

McGlathery K, Marino JR & Howarth RW (1994) Variable rates of phosphate uptake by shallow marine sediments: Mechanisms and ecological significance. Biogeochemistry 25: 127–146

McGlathery KJ, Howarth RW & Marino R (1992) Nutrient limitation of the macroalga, *Penicillus capitatus*, associated with subtropical seagrass meadows in Bermuda. Estuaries 15: 18–25

Meybeck M (1993) C, N, P, and S in rivers: from sources to global inputs. In: Wollast R Mackenzie FT & Chou L (Eds) Interaction of C, N, P, and S. Biogeochemical Cycles and Global Change (pp 163–193). Springer-Verlag, Berlin, Germany

Moran MA & Hodson RE (1994) Dissolved humic substances of vascular plant origin in a coastal marine environment. Limnol. Oceanog. 39: 762–771

Morell JM & Corredor JE (1993) Sediment nitrogen trapping in a mangrove lagoon. Est. Coast. & Shelf Sci. 37(2): 203–212

Morell JM & Corredor JE (In Review) Fertilization of the Eastern Caribbean by the Orinoco River: How does it spread so far? Submitted to J. Geophys. Res.

Morese JW, Zullig JJ, Bernstein LD, Millero FJ, Milne P, Mucci A & Choppin GR (1985) Chemistry of calcium carbonate-rich shallow water sediments in the Bahamas. Am. J. Sci. 285: 147–185

Morse JW, Zullig JJ, Iverson RL, Choppin GR, Mucci A & Millero FJ (1987) The influence of seagrass beds on carbonate sediments in the Bahamas. Mar. Chem. 22: 71–83

Mosquera AI, Corredor JE & Morell JM (1998) Exponential collapse of benthic depurative capacity in a eutrophic tropical marine ecosystem: A threshold response to organic nitrogen loading. Chemistry and Ecology 14: 341–355

Muller Karger FE, McClain CR, Fisher TR, Esaias WE & Varela R (1989) Pigment distribution in the Caribbean Sea: Observations from space. Prog. Oceanog. 23: 23–64

Naqvi SWA & Noronha RJ (1991) Nitrous oxide in the Arabian Sea. Deep-Sea Res. 38: 872–890

Nevison CD, Weiss RF & Ericksson DJ III (1995) Global oceanic emissions of nitrous oxide. J. Geophys. Res. 100 (C8): 15,809–15,820

Nieves FY & Corredor JE (1987) Gradientes de fijación de nitrógeno en los sedimentos marinos asociados a arrecifes coralinos en el suroeste de Puerto Rico. An. Inst. Inv. Mar. Punta de Betín. 17: 27–37

Nixon SW, Ammerman JW, Atkinson LP, Berounsky VM, Billen G, Boicourt WC, Boynton WR, Church TM, DiToro DM, Elmgren R, Garber JH, Gibline AE, Jahnke RA, Owens NJP, Pilson MEQ & Seitzinger SP (1996) The fate of nitrogen and phosphorus at the land-sea margin of the North Atlantic Ocean. Biogeochemistry 35: 141–180

Onuf C, Teal J & Valiela I (1977) The interactions of nutrients, plant growth, and herbivory in a mangrove ecosystem. Ecology 58: 514–526

Pelegri SP, Rivera-Monroy VH & Twilley RR (1997) A comparison of nitrogen fixation (acetylene reduction) among three species of mangrove litter, sediments, and pneumatophores in south Florida, U.S.A. Hydrobiologia 356: 73–79

Pelegri SP & Twilley RR (1998) Interactions between nitrogen fixation (acetylene reduction) and leaf litter decomposition of two mangrove species from south Florida, U.S.A.: potential inhibitory effects of phenolics. Marine Biology 131: 53–61

Potts M (1979) Nitrogen fixation (acetylene reduction) associated with communities of heterocystous and non-heterocystous blue-green algae on mangrove forests of Sinai. Oecologia 39: 359–373

Richardson M (1985) Nitrification Inhibition in the Treatment of Sewage. The Royal Society of Chemistry, Whitstable

Rivera-Monroy VH & Twilley RR (1996) The relative role of denitrification and immobilization in the fate of inorganic nitrogen in mangrove sediments (Terminos Lagoon, Mexico). Limnol. Oceanog. 41: 284–296

Rivera-Monroy VH, Day JW, Twilley RR, Vera-Herrera F & Coronado-Molina C (1995a) Flux of nitrogen and sediment in a fringe mangrove forest in Terminos Lagoon, Mexico. Estuarine, Coastal and Shelf Science 40: 139–160

Rivera-Monroy VH, Twilley RR, Boustany RG, Day JW, Vera-Herrera F & Ramirez MC (1995) Direct denitrification in mangrove sediments in Terminos Lagoon, Mexico. Marine Ecology Progress Series 126: 97–109

Rodhe H (1990) A comparison of the contribution of various gases to the green-house effect. Science 248: 1217–1219

Ryther JH, Menzel DW & Corwin N (1967) Influence of the Amazon River outflow on the ecology of the Western Tropical Atlantic I. Hydrography and nutrient chemistry. J. Mar. Res. 25: 69–83

Sánchez-Suárez IG, Troncone-Osorio FC & Díaz-Ramos JR (1995) The phytoplankton form the Gulf of Paria, Venezuela (June 1984) Marine Biology Acta Científica Venezolana 46: 192–205

Seitzinger SP (1988) Denitrification in freshwater and coastal marine ecosystems: ecological and geochemical significance Limnol. Oceanogr. 33(4), pt 2: 702–724

Short FT, Dennison WC & Capone DC (1990) Phosphorus-limited growth of the tropical seagrass *Syringodium filiforme* in carbonate sediments. Mar. Ecol. Prog. Ser. 62: 169–174

Smith SV (1984) Phosphorus vs nitrogen limitation in the marine environment. Limnol. Oceanogr. 29: 1149–1160

Twilley RR (1997) Mangrove wetlands. In: Messina M & Connor W (Eds) Southern Forested Wetlands: Ecology and Management (pp 445–473). CRC Press, Boca Raton, Florida, U.S.A.

Twilley RR, Pozo M, Garcia VH, Rivera-Monroy VH, Zambrano R & Bodero A (1997) Litter dynamics in riverine mangrove forests in the Guayas River estuary, Ecuador. Oecologia 111: 109–122

Twilley RR (1995) Properties of mangrove ecosystems related to the energy signature of coastal environments. In: Hall CAS (Ed.) Maximum Power: The Ideas and Applications of Odum HT (pp 43–62). University Press of Colorado, Niwot, CO, U.S.A.

Twilley RR, Lugo AE & Patterson-Zucca C (1986) Production, standing crop, and decomposition of litter in basin mangrove forests in southwest Florida. Ecology 67: 670–683

Tyrrell T & Law CS (1997) Low nitrate:phosphate ratios in the global ocean. Nature 387: 793–796

Van der Valk AG & Attiwill PM (1984) Acetylene reduction in an *Avicennia marina* community in southern Australia. Australian J. Botany 32: 157–164

Vanzella A, Guerrero MA & Jones RD (1989) Effect of CO and light on ammonium and nitrite oxidation by chemolithotrophic bacteria. Mar. Ecol. Prog. Ser. 57: 69–76

Vitousek PM & Howarth RW (1991) Nitrogen limitation on land and in the sea: How can it occur? Biogeochem. 13: 87–115

Weiss RF (1981) The temporal and spatial distribution of atmospheric nitrous oxide. J. Geophys. Res. 86: 7185–7195

Yoshioka P, Owen G & Pesante D (1985) Spatial and temporal variations in Caribbean zooplankton near Puerto Rico. J. Plankton Res. 7: 733–751

Zuberer DA & Silver WS (1978) Biological nitrogen fixation (acetylene reduction) associated with Florida mangroves. Applied and Environmental Microbiology 35: 567–575

Biogeochemistry **46:** 179–202, 1999.

Ecosystem constraints to symbiotic nitrogen fixers: a simple model and its implications

PETER M. VITOUSEK[1] & CHRISTOPHER B. FIELD[2]
[1]*Department of Biological Sciences, Stanford University, Stanford, California 94305 U.S.A.;* [2]*Department of Plant Biology, Carnegie Institution of Washington, Stanford, California 94305 U.S.A.*

Received 10 December 1998

Key words: carbon dioxide, grazing, nitrogen fixation, nitrogen limitation, phosphorus, shade

Abstract. The widespread occurrence of *N* limitation to net primary production (NPP) and other ecosystem processes, despite the ubiquitous occurrence of *N*-fixing symbioses, remains a significant puzzle in terrestrial ecology. We describe a simple simulation model for an ecosystem containing a generic nonfixer and a symbiotic *N* fixer, based on: (1) a higher cost for *N* acquisition by *N* fixers than nonfixers; (2) growth of fixers and fixation of *N* only when low *N* availability limits the growth of nonfixers, and other resources are available; and (3) losses of fixed *N* from the system only when the quantity of available *N* exceeds plant and microbial demands. Despite the disadvantages faced by the *N* fixer under these conditions, *N* fixation and loss adjust *N* availability close to the availability of other resources, and biomass and NPP in this simple model can be substantially but only transiently *N* limited. We then modify the model by adding: (1) losses of *N* in forms other than excess available *N* (e.g., dissolved organic *N*, trace gases produced by nitrification); and (2) constraints to the growth and activity of *N* fixers imposed by differential effects of shading, *P* limitation, and grazing. The combination of these processes is sufficient to describe an open system, with input from both precipitation and *N* fixation, that is nevertheless strongly *N*-limited at equilibrium. This model is useful for exploring causes and consequences of constraints to *N* fixation, and hence of *N* limitation, and we believe it will also be useful for evaluating how *N* fixation and limitation interact with elevated CO_2 and other components of global enviromental change.

Introduction

The supply of nitrogen often limits the growth of plants, the composition of communities, the productivity of ecosystems, and other population, community, and ecosystem processes. Fertilization experiments suggest that added *N* enhances productivity in many or most intensive agricultural systems, temperate forests, temperate and tropical grasslands, boreal forests, and arctic/alpine tundras (Shaver & Chapin 1980; Miller 1981; Lee et al. 1983; Van Bremen & de Wit 1983; Hunt et al. 1988; Bonan 1990; Bowman et al. 1993; Magill et al. 1997). Correlative studies also show a tight connection between *N* mineralization and forest productivity (Reich et al. 1997), and ecosystem models suggest *N* limitation is widespread (McGuire et al. 1992; Parton et al. 1993). Perhaps the most compelling evidence for the importance of *N* limitation is the tens of billions of dollars that humanity spends annually on *N* fertilizer.

However, despite all of the sustained scientific effort that has gone into analyzing the cycle of *N*, there are no generally accepted answers to the questions: Why does *N* supply limit productivity in so many natural and managed ecosystems? *N*-fixers should have a substantial competitive advantage in *N*-limited systems, and as a byproduct of their activity they should increase the quantity and availability of *N* in the system as a whole. Why don't they do so? In a world where biological *N*-fixation is ubiquitous, how can *N* limitation be widespread? What keeps *N* fixers from adding sufficient *N* to terrestrial ecosystems to bring the supply of *N* more or less into equilibrium with other potentially limiting resources?

One credible set of answers is that in the long term, *N* fixation does adjust *N* supply close to the availability of other resources (Schimel et al. 1997). In this case, *N* limitation would be a transient and/or marginal phenomenon – as indeed we know it is in most temperate lakes. Outside of frequently and intensely disturbed systems that are actively maintained in a transient state, such as annual crops or fire-dominated areas (Seastedt et al. 1991), *N* limitation might be more apparent than real, more proximate than ultimate. Alternatively, there are a number of mechanisms that could prevent *N* fixers from responding to even profound *N* limitation, and could thereby maintain *N* limitation in the long term. These mechanisms include energetic constraints to growth of or colonization by *N* fixers, disproportionate *P* (or other element) limitation to fixers as opposed to nonfixers, and disproportionately high rates of herbivory on fixers as opposed to nonfixers (Vitousek & Howarth 1991).

Determining whether *N* limitation is transient or sustained, marginal or substantial, is fundamental to our understanding of the biogeochemistry of terrestrial ecosystems (Vitousek & Howarth 1991). It also represents a major uncertainty in our ability to predict the responses of terrestrial ecosystems to components of global environmental change. For example, humanity has greatly increased the fixation of *N*, and its deposition to terrestrial ecosystems (Galloway et al. 1995; Vitousek et al. 1997a). Where that fixed *N* reaches *N*-limited systems, it can alter their composition, dynamics and *C* storage profoundly (Berendse et al. 1993; Schimel et al. 1996; Howarth et al. 1996; Nixon et al. 1996); where recipient systems are not *N*-limited, or *N* limitation is marginal and/or short-lived, its consequences are smaller and less certain (Matson et al., this volume).

Another example – much of the uncertainty about long-term ecosystem responses to elevated atmospheric CO_2 is related to N limitation. The stimulation of plant growth by elevated CO_2 in short-term experiments is highly variable, but median increases are 30–40% (Poorter 1993; Poorter et al. 1996; Curtis 1996; Koch & Mooney 1996), substantially above the 5–10% increases in long-term NPP predicted by biogeochemical models (VEMAP 1995; Melillo et al. 1996). N limitation exerts profound constraints to CO_2 responses in most biogeochemical models (McMurtrie & Comins 1996; Rastetter et al. 1997), even though the empirical evidence for decreased CO_2 sensitivity under N limitation is mixed. Idso and Idso (1994) and Lloyd and Farquhar (1996) found no consistent effect of nutrient status on response to elevated CO_2, though a number of experiments tie the low sensitivity of particular plants or ecosystems to nutrient effects (e.g. Larigauderie et al. 1988; Diaz et al. 1993; Oechel et al. 1994; Leadley & Körner 1996).

What accounts for this contrast in results and perspective? One component involves species characteristics; species from nutrient-poor sites show small sensitivity to either nutrient (Chapin 1980; Chapin et al. 1986; Field et al. 1992) or CO_2 additions (Poorter 1993; Poorter et al. 1996; Hunt et al. 1991, 1993). Another component involves time scale. Most growth experiments last from less than one to a few years, while in the biogeochemistry models, the important feedbacks involving nutrient limitation emerge only after several years, or even several decades, of CO_2 exposure (McMurtrie & Comins 1996). N limitation gradually intensifies with the immobilization of N in plant biomass, litter, and soil organic matter, even where growth at elevated CO_2 has no depressive effect on tissue decomposability (O'Neill & Norby 1996; Franck et al. 1996).

Alternatively, it is possible that in the medium to long term, increased CO_2 could stimulate N fixation on land, making any N constraint on plant response to increased CO_2 a transient phenomenon (Gifford 1992). The growth response to elevated CO_2 appears to be somewhat larger in legumes than in other species (Hunt et al. 1991, 1993; Poorter 1993), and elevated CO_2 often enhances symbiotic fixation under controlled conditions – usually as a consequence of increased legume growth, rather than increased N fixation per unit of legume biomass (Phillips et al. 1976; Finn & Brun 1982; Norby 1987; Arnone & Gordon 1990; Thomas et al. 1991; Ryle et al. 1992). What are the factors that regulate changes in N fixation under elevated CO_2? Will the stimulation of fixation by planted, weeded, tended, often fertilized N fixers observed in experiments carry over to increased rates of N fixation in natural ecosystems?

In this paper, we develop a conceptual model for the regulation of symbiotic *N* fixation in terrestrial ecosystems. We then use this model and a set of simple simulations to evaluate several mechanisms that could constrain rates of *N* fixation, and could thereby allow *N* limitation to persist indefinitely. Finally, we explore how these mechanisms could interact with components of human-caused global environmental change.

A conceptual model

The basic assumptions of our model include: (1) The energetic cost of acquiring *N* through biological fixation exceeds the cost of acquiring *N* from the soil, whenever there is more than a trace of available *N* in the soil; (2) Symbiotic *N* fixers require and maintain higher concentrations of *N* and *P* in their tissue than do nonfixers; and (3) Symbiotic *N* fixers only acquire *N* through fixation, rather than obtaining it from the soil. The first two of these are readily supported (Pate 1986; Gutschick 1987; Smith 1992; McKey 1994). The last is not correct – symbiotic fixers can and do take *N* from soil, when it is available – but it simplifies the model substantially. More importantly, our purpose is to evaluate the controls of *N* fixation and the plant growth supported by it, not to model the growth of legumes per se (at least not yet).

We begin development of the conceptual model by assuming that for a nonfixer, the cost of acquiring soil *N* is a function of *N* availability, as illustrated in Figure 1. The x-axis represents the total quantity of *N* that could be available in soil annually (net *N* mineralization, inputs via precipitation, any other sources of available *N*); the quantity of *N* that actually becomes available in a particular system is indicated by point *A* on the figure. Some of this available *N* can be acquired at a relatively low (and fixed) energetic cost, one that reflects the enzymatic machinery involved in *N* uptake and transport. However, after half of the available *N* has been taken up, obtaining the remainder requires increasing investment in roots, or carriers, or other mechanisms by which *N* can be obtained from dilute concentrations in soil. Costs increase to the point where essentially all of the available *N* in soil has been taken up; after that point, further investment in acquiring soil *N* is unrewarding. We model the cost function as increasing linearly, although in fact it probably increases exponentially.

In contrast, we assume that the cost to a symbiotic *N* fixer of acquiring a unit of N_2 is constant, independent of the quantity of *N* fixed (Figure 1). We assume that the two cost functions cross at point *A*, where essentially all of the fixed *N* in the soil has been utilized. To the left of that point, nonfixers have a lower cost for *N* acquisition than fixers and can (we assume) outcom-

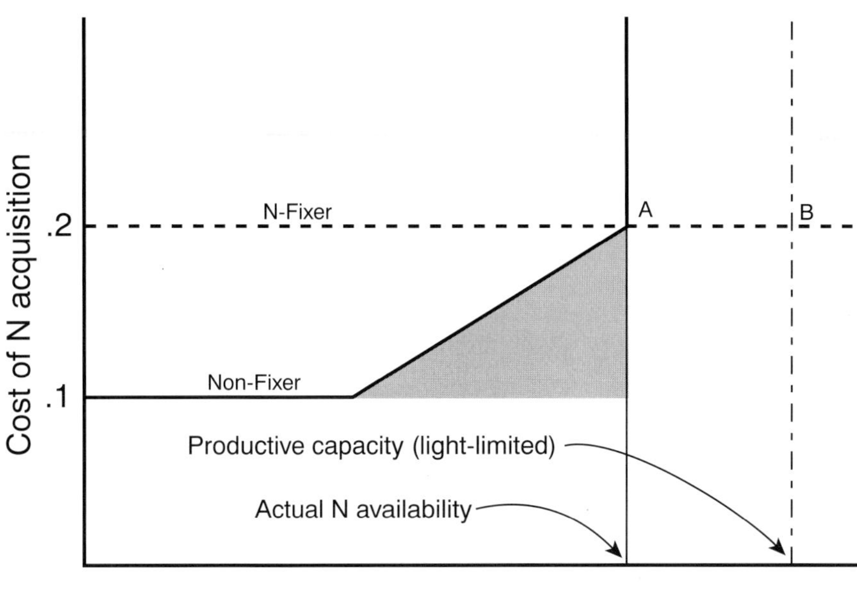

Figure 1. A conceptual model for the costs of N acquisition by nonfixers and N fixers. The x-axis represents N availability, as quantity of N/area (e.g. kg $\cdot$ ha^{-1} $\cdot$ yr^{-1}). Any particular ecosystem will have a light-limited productive capacity, and a level of N availability required to support that productivity (B); any system will also have an actual level of N availability (A), which in this example is below the productive capacity. The dark solid line represents the cost of N acquisition by a nonfixer; we assume that N acquisition is relatively inexpensive when a large quantity of available N remains in the system, but the cost increases as a greater proportion of the available N is utilized. As the available N in the system approaches full utilization, the cost increases very steeply. In contrast, N fixers (dark dashed line) have a relatively high but constant cost of N acquisition. In this particular example, some light would remain unutilized after all of the available N had been taken up (A), and N fixers could use that light to grow, fix N, and (ultimately) increase the N availability in the system towards B. The shaded area under the nonfixer's cost curve represents energy that could be allocated for purposes other than N acquisition, if more N were available.

pete them, while to the right of point A, fixers can obtain N and nonfixers cannot. Finally, we assume that there is a productive capacity for each site, set by the availability of resources other than N. Initially, we will use light as that ultimate resource. Point B (Figure 1) then corresponds to the quantity of available N that would be required to support nonfixers at this maximum (light-limited) level of productivity.

Given this simple formulation, the availability of N in an ecosystem should adjust to that of other limiting resources. If point A is to the left of B as in Figure 1, then symbiotic N fixers should grow, fix N, and gradually increase N supply until its availability is equivalent to B, and fixers no longer have an advantage. At this point, N supply could still limit production of nonfixers, in that large additions of N would allow plants to reallocate some of the resources that they had used to acquire N. However, only energy equivalent to the shaded area in Figure 1 would be available for reallocation, so any increase in production would be marginal.

Simulation

Based on the conceptual model in Figure 1, we put together a simple simulation of production and N cycling in an ecosystem containing a nonfixer and a symbiotic N-fixer. Nitrogen availability is calculated as the sum of net N mineralization, N inputs in precipitation or fertilizer, and any carryover from the previous year. The potential N uptake by nonfixing plants is set to 100 units of N (say 100 kg/ha/yr), corresponding to a maximum potential NPP of 5000 units of C (minus the C costs of N acquisition). If more than 100 units of N are available, the quantity in excess of 100 units is not taken up; if less than 100 units are available, all are taken up by nonfixers. Production of the nonfixers (NPPNF) is then calculated as

$$\text{NPPNF} = CN\text{nf} * N\text{uptake} * (1 - \text{costNF}), \tag{1}$$

where CNnf is the C/N ratio (uncorrected for cost of N uptake) of nonfixers, Nuptake is uptake of N by nonfixers, and costNF is the C expended in acquiring N. costNF is determined using the function in Figure 1, in which the cost is 0.1 up to half of total N availability, then increases linearly to 0.2 for the last increment of available N.

Production by symbiotic N fixers (NPPfix) is then calculated as

$$\text{NPPfix} = 5000 * (1 - \text{fixCOST}) * (4500 - \text{NPPNF})/4500, \tag{2}$$

where fixCOST is the cost of fixing N (which we set equal to 0.3, calculated as 0.2 per unit of N, in terms comparable to nonfixers, multiplied by 1.5 to account for the greater N concentration in fixers). The potential productivity (independent of the cost of acquiring N) is 5000; the light available for growth by fixers, after nonfixers have taken what they can, is proportional to $(4500 - \text{NPPNF})/4500$, 4500 being the maximum NPP of nonfixers that can be realized in practice. In effect, this formulation gives nonfixers priority for light in proportion to the soil N they can acquire. If there is still light available after nonfixers have taken what they can, then N fixers can grow (and fix N). Water and other resources could be treated similarly.

The production of litter C and N by fixers and nonfixers (independently) is set to 10% annually of the C and N in plant biomass. The rate of decomposition is set to 5% of soil C per year, and net N mineralization (Nmin) is calculated as

$$N\text{min} = \text{soil}N - \text{soil}C/CN\text{crit}, \tag{3}$$

where soil$N$ and soil$C$ are the quantities of organic N and C (respectively) in soil, and CNcrit is the critical ratio of C to N, above which N is immobilized by microbes and below which N is mineralized.

The simulation maintains mass balance of C and N in plants, soils, and the system as a whole; it also constrains potential production and N uptake by fixers and nonfixers when their initial biomass is low. We include a relatively low rate of input of fixed N in precipitation (2 units per year), one designed to represent unpolluted regions. In this formulation, outputs of N occur only when N remains in the available pool after all biological demands are met. A complete listing of the simulation in MATLAB (including the extensions described in subsequent sections) is available from the authors at http://jasper.stanford.edu/chrisweb/flab/flab.html.

Initial results

We used this simple model to simulate the development of an ecosystem on a new substrate, one that starts with no organic C or N in soil. Initially, very low levels of N are available in soil; consequently the growth of nonfixers is limited by N, light is available, and N fixers dominate the site and add substantial quantities of N to the system (Figures 2a, 3a). As this fixed N enters the soil, rates of N mineralization eventually increase, and nonfixers colonize, grow, and ultimately use almost all the available light, replacing the fixers. Similar patterns of species replacement are observed in the field during primary successions in which symbiotic N fixers are present (Walker & Syers 1976; Van Cleve et al. 1991; Chapin et al. 1994). Approximately 150 years into the simulation, the system reaches equilibrium levels of production, biomass, N mineralization, etc; N fixers are then very sparse in the system. Nevertheless, N remains limiting to nonfixers, if only marginally. We simulate additions of N to the equilibrium system by adding 50 units of inorganic N a year for a 20-year period beginning in year 300, and find a small increase in biomass of nonfixers that reflects their lower cost of acquiring N when its availability is high (Figures 1, 3a).

We also evaluated ecosystem development without N fixers, forcing the system to draw upon inputs of fixed N from the atmosphere. Under these

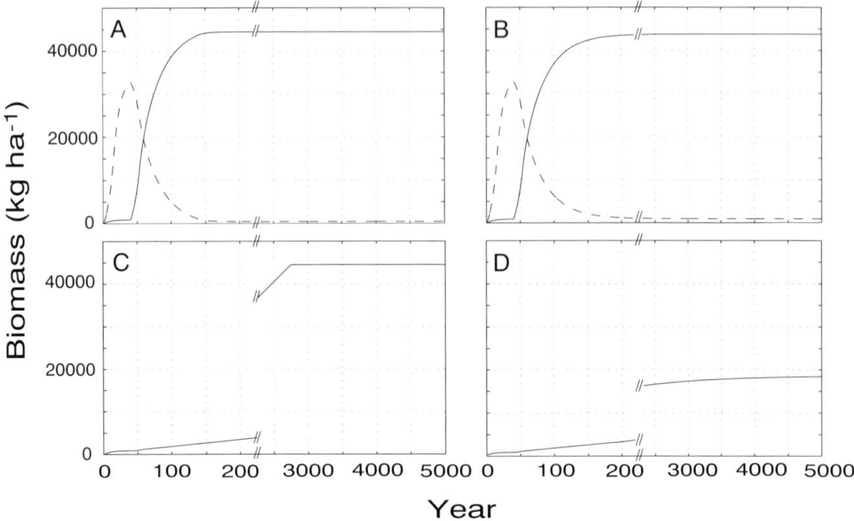

Figure 2. Simulated biomass of nonfixers (solid lines) and N fixers (dashed lines) through long-term ecosystem development on a site that begins with no C, N, or available P in soil. The panels evaluate combinations of the presence or absence of N fixers and differing pathways of N loss. Those on top (a, b) include N fixers while those on the bottom (c, d) exclude them; those on the left (a, c) include only losses of N from the pool of available N that remains in the soil after plant and microbial uptake, while those on the right (b, d) also include losses of 5% of mineralized N. Note the change in scale on the x-axes.

conditions, the system eventually reaches the same equilibrium level of production, etc. – as indeed it must, as long as there are any N inputs, and only available N in excess of biological requirements is lost. However, it takes much longer to reach that equilibrium (2700 versus 150 years) (Figure 2c). Moreover, if we simulate N losses occurring by pathways other than removal of excess available N, by removing 5% of mineralized N annually, then in the absence of N fixers the system equilibrates at a lower biomass, productivity, N mineralization, etc., as observed in the simpler model of Vitousek et al. (1998) (Figure 2d). In practice, leaching of dissolved organic N and emissions of N trace gases produced during nitrification (though not denitrification) could represent losses that are independent of the existence of excess available N (Hedin et al. 1995; Parton et al. 1996). Without N fixers, such N losses can allow substantial N limitation to persist at equilibrium (Vitousek et al. 1998).

Finally, we evaluated the consequences of a 5% loss of mineralized N in the presence of N fixers. In this scenario, fixers remain within the system at equilibrium, though at a relatively low level (Figure 2b, 3b). Their activity is sufficient to offset losses of N, and (together with atmospheric inputs), to

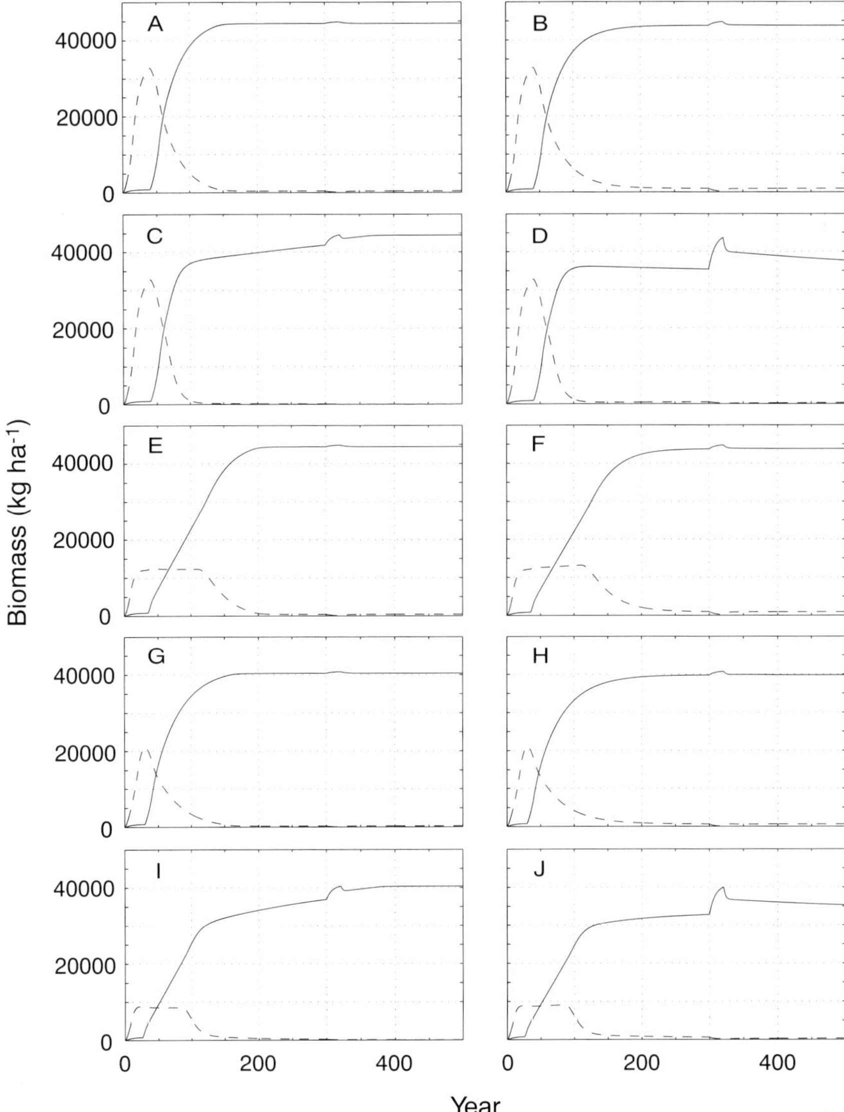

Figure 3. Simulated biomass of nonfixers (solid lines) and fixers (dashed lines); the panels evaluate combinations of pathways of N loss and constraints on N fixers. Those on the left (a, c, e, g, i) include losses of N only from the pool of available N that remains in the soil after microbial and plant uptake; the panels on the right (b, d, f, h, j) also include losses of 5% of mineralized N. After 300 y, 50 units of N/year are added for 20 years, to evaluate the strength of N limitation. (a, b) The basic model, as outlined in the text. (c, d) Inclusion of a shade limitation to the colonization and growth of N fixers. (e, f) Inclusion of phosphorus limitation. P supply can constrain both nonfixers and fixers, but N fixers require more P and so are more severely constrained. (g, h) Inclusion of grazing, which removes 1% of the biomass of nonfixers and 5% of the biomass of N fixers annually. (i, j) The combination of all three constraints to the growth of N fixers.

less P is available, then the entire minP pool is taken up, and N uptake is constrained to the level determined by P uptake. NPP of the nonfixers is then calculated as in Equation (1).

We calculate the potential production by fixers, as it is constrained by P supply (potPNPPfix), as

$$\text{potPNPPfix} = \text{minP} * CP\text{f}, \tag{5}$$

where minP is now the quantity of available P remaining after uptake by nonfixers, and CPf is the carbon to phosphorus ratio of fixers. This potential production is compared to that calculated by Equation (2) above, and the smaller of the two potential productivities is used.

The production of litter P is set to 10% of biomass P, in parallel with C and N, and P mineralization is calculated analogously to N mineralization (Equation (3)). To account for the much lower mobility of P than N, we assume that only 10% of the available P remaining in the soil after biological demands are met can be lost from the system annually. This loss could represent leaching of inorganic P, or it could represent the formation of occluded P (cf Walker & Syers 1976). Twenty percent of the available P remaining after uptake is carried over to the next year, and 70% goes into a labile inorganic pool with a turnover time of 10 years. We assume P inputs via weathering to be 2 units/year, with a brief pulse of more rapid weathering in the initial stage of soil development.

The simulated consequences of P limitation are summarized in Figures 3(e) and 3(f). Pathways of N loss have little effect on the responses of biomass or other ecosystem characteristics to P limitation. Compared to the simple model (Figures 3(a, b)), a low supply of P strongly constrains the growth of N fixers early in ecosystem development, extends the period over which fixers are active by about 60 years, and delays equilibrium biomass in the system as a whole by about 50 years. Early in development, the production of nonfixers is limited by N, while that of fixers is limited by P. Additions of N would stimulate production by nonfixers, up to the point where they too would become P-limited. Additions of P would stimulate N fixers, which in turn would add N and stimulate nonfixers, bringing the whole system to equilibrium more rapidly. In this case, what appears as N limitation (to nonfixers) could actually be P limitation in disguise (Vitousek & Howarth 1991).

Later in development, sufficient P has weathered from the substrate so that P no longer constrains N fixation (in this model). However, this simulation only considers losses of excess available P, the inorganic P that remains after biological demands are met. The combination of a much lower rate of P weathering in very old soils (Walker & Syers 1976; Crews et al. 1995;

Newman 1995; Vitousek et al. 1997b) with losses of P via other pathways (such as leaching of dissolved organic P) could in practice cause P limitation to N fixation in old as well as very young soils.

A low supply of elements other than P could also constrain N fixation, if N fixers require more of a particular element than do nonfixers. Molybdenum and iron are perhaps the most interesting of these other elements; both are necessary to the functioning of the nitrogenase enzyme, and Mo in particular is not even required by nonfixers that use ammonium as their sole N source. There is some evidence from both marine and terrestrial ecosystems that additions of Mo sometimes can stimulate rates of N fixation (Howarth & Cole 1985; Silvester 1989).

Control by grazing

The growth of many grazing animals is limited by access to protein more than it is by access to energy. N fixers are systematically richer in protein than nonfixers, and are often preferred by both vertebrate and invertebrate herbivores (Hulme 1994, 1996). Selective grazing on N fixers could suppress N fixation and thereby keep N availability from equilibrating with the supply of other resources, as Ritchie and Tilman (1995) demonstrated in old-fields and savannas at Cedar Creek, Minnesota. Alternatively, N fixers could maintain more chemical defenses against grazers than nonfixers, reducing losses to grazing at the expense of effectively raising the cost of N fixation.

We simulated the effects of grazing by defining a fraction of the nonfixers' biomass that is consumed annually (grazefrac). We established a higher grazing pressure on N fixers using a multiple of this fraction. For the runs here, we set grazing on the nonfixers at 1% and grazing on fixers at 5%. We assume that all of the N and P, and half of the C, in grazed plant tissue is transferred to the soil organic pool; the remainder of the C is respired by grazers.

The effects of differential grazing on fixers versus nonfixers are summarized in Figures 3(g) and 3(h); pathways of N loss have little effect on responses to grazing. Relative to the control case (Figures 2(a, b)), grazing reduces the peak biomass of N fixers by about 35%; it thereby delays the whole system's approach to equilibrium by about 20 years. The equilibrium biomass of the nonfixer is reduced slightly by the direct effects of grazing. More interestingly, because grazers return material with lower C:N and C:P ratios to the soil than does litterfall, grazing causes substantial net mineralization of N from soil organic matter, and consequently increased growth of the nonfixer, to begin almost 10 years earlier than in the control case.

Lee JA, Harmer R & Ignaciuk R (1983) Nitrogen as a limiting factor in plant communities. In: Lee JA, McNeill S & Rorison IH (Eds) Nitrogen as an Ecological Factor (pp 95–112). Blackwell Scientific, Oxford

Lloyd J & Farquhar GD (1996) The CO_2 dependence of photosynthesis, plant growth responses to elevated atmospheric CO_2 concentrations and their interaction with soil nutrient status. I. General principles and forest ecosystems. Funct. Ecology 10: 4–33

Magill AH, Aber JD, Hendricks JJ, Bowden RD, Melillo JM & Steudler P (1997) Biogeochemical response of forest ecosystems to simulated chronic nitrogen deposition. Ecol. Applic. 7: 402–415

Martinelli LA, Piccolo MC, Townsend AR & Vitousek PM. Nitrogen stable isotope composition of leaves and soil: tropical versus temperate forests. Biogeochem., this volume

Matson PA, McDowell WH, Townsend AR & Vitousek PM. The globalization of N deposition: ecosystem-level consequences. Biogeochem., this volume

McGuire AD, Melillo JM, Joyce LA, Kicklighter DW, Grace AL, Moore B III & Vorosmarty CJ (1992) Interaction between carbon and nitrogen dynamics in estimating net primary productivity for potential vegetation in North America. Glob. Biogeochem. Cyc. 6: 101–124

McKey D (1994) Legumes and nitrogen: the evolutionary ecology of a nitrogen-demanding lifestyle. In: Sprent JL & McKey D (Eds) Advances in Legume Systematics: Part 5 – The Nitrogen Factor (pp 211–228). Royal Botanic Gardens, Kew, England

McMurtrie RE & Comins HN (1996) The temporal response of forest ecosystems to doubled atmospheric CO_2 concentration. Glob. Change Biol. 2: 49–59

Melillo JM, Prentice IC, Farquhar GD, Schulze E-D & Sala OE (1996) Terrestrial ecosystems: Biotic feedbacks to climate. In: Houghton J, Meira LG Filho, Callander BA, Harris N, Kattenberg A & Maskell K (Eds) Intergovernmental Panel on Climate Change 1995: Scientific Assessment of Climate Change (pp 447–481). Cambridge University Press, Cambridge

Miller HG (1981) Forest fertilization: some guiding concepts. Forestry 54: 157–167

Newman EI (1995) Phosphorus inputs to terrestrial ecosystems. J. Ecology 83: 713–726

Nixon SW, Ammerman JW, Atkinson LP, Berounsky VM, Billen G, Boicourt WC, Boyton WR, Church TM, Ditoro DM, Elmgren R, Garber JH, Giblin AE, Jahnke RA, Owens NPJ, Pilson MEQ & Seitzinger SP (1996) The fate of nitrogen and phosphorus at the land-sea margin of the North Atlantic Ocean. Biogeochem. 35: 141–180

O'Connell AM (1994) Decomposition and nutrient content of litter in a fertilized eucalypt forest. Biol. Fert. Soils 17: 159–166

Oechel WC, Cowles S, Grulke N, Hastings SJ, Lawrence B, Prudhomme T, Riechers G, Strain B, Tissue D & Vourlitis G (1994) Transient nature of CO_2 fertilization in Arctic tundra. Nature 371: 500–503

Ojima DS, Schimel DS, Parton WJ & Owensby CE (1994) Long- and short-term effects of fire on nitrogen cycling in tallgrass prairie. Biogeochem. 24: 67–84.

O'Neill EG & Norby RJ (1996) Litter quality and decomposition rates of foliar litter produced under CO_2 enrichment. In: Koch GW & Mooney HA (Eds) Carbon dioxide and terrestrial ecosystems (pp 87–104). Academic Press, San Diego, U.S.A.

Parton WJ, Scurlock JMD, Ojima DS, Gilmanov TG, Scholes RJ, Schimel DS, Kirchner T, Menaut J-C, Seastedt T, Garcia Moya E, Kamnalrut A & Kinyamarie JI (1994) Observations and modelling of biomass and soil organic matter dynamics for the grassland biome worldwide. Glob. Biogeochem. Cyc. 7: 785–809

Parton WJ, Mosier AR, Ojima DS, Valentine DW, Schimel DS, Weier K & Kulmala AE (1996) Generalized model for N_2 and N_2O production from nitrification and denitrification. Glob. Biogeochem. Cyc. 10: 401–412

Pate JS (1986) Economy of symbiotic N fixation. In: Givnish TJ (Ed) On the Economy of Plant Form and Function (pp 299–325). Cambridge University Press, Cambridge, U.K.

Phillips DA, Newell KD, Hassell SA & Felling CE (1976) The effect of CO_2 enrichment on root nodule development and symbiotic N_2 reduction in *Pisum satiuum*. L. Am. J. Botany 63: 356–362

Poorter H (1994) Interspecific variation in the growth response of plants to an elevated ambient CO_2 concentration. Vegetatio 104/105: 77–97

Poorter H, Roumet C & Campbell BD (1996) Interspecific variation in the growth response of plants to elevated CO_2: A search for functional types. In: Körner C & Bazzaz FA (Eds) Carbon Dioxide, Populations & Communities (pp 375–412). Academic Press, San Diego, U.S.A.

Prescott CE, McDonald MA, Gessel SP & Kimmins JP (1994) Long-term effects of sewage sludge and inorganic fertilizers on nutrient turnover in litter in a coastal Douglas fir forest. Forest Ecol. Manage. 59: 149–164

Reich PB, Grigal DF, Aber JD & Gower ST (1997) Nitrogen mineralization and productivity in tree stands on diverse soils. Ecology 78: 335–347

Rastetter EB, Agren GI & Shaver GR (1996) Responses to increased CO_2 concentration in N-limited ecosystems: application of a balancing-nutrition, coupled-element-cycles model. Ecol. Applic. 7: 444–460

Ritchie ME & Tilman D (1995) Responses of legumes to herbivores and nutrients during succession on a nitrogen-poor soil. Ecology 76: 2648–2655

Ryle GJA, Powell CE & Davidson IA (1992) Growth of white clover, dependent on N_2 fixation, in elevated CO_2 and temperature. Ann. Botany 70: 221–228

Schimel DS, Braswell BH, McKeown R, Ojima DS, Parton WJ & Pulliam W (1996) Climate and nitrogen controls on the geography and time-scales of terrestrial biogeochemical cycling. Glob. Biogeochem. Cyc. 10: 677–692

Schimel DS, Brassell BH & Parton WJ (1997) Equilibration of the terrestrial water, nitrogen, and carbon cycles. PNAS 94: 8280–8283

Schindler DW (1977) Evolution of phosphorus limitation in lakes. Science 195: 260–262

Seastedt TR, Briggs JM & Gibson DJ (1991) Controls of nitrogen limitation in tallgrass prairie. Oecol. 87: 72–79

Shaver GR & Chapin FS III (1980) Response to fertilization by various plant growth forms in an Alaskan tundra: nutrient accumulation and growth. Ecology 61: 662–675

Silvester WB (1989) Molybdenum limitation of asymbiotic nitrogen fixation in forests of Pacific Northwest America. Soil Biol. Biochem. 21: 283–289

Smith VH (1992) Effects of nitrogen:phosphorus supply ratios in nitrogen fixation in agricultural and pastoral systems. Biogeochem. 18: 19–35

Thomas RB, Richter DD, Ye H, Heine PR & Strain BR (1991) Nitrogen dynamics and growth of seedlings of an N-fixing tree (*Gliricidia sepium* (Jacq.) Walp.) exposed to elevated atmospheric carbon dioxide. Oecol. 88: 415–421

van Bremen H & de Wit CT (1983) Rangeland productivity and exploitation in the Sahel. Science 221: 1341–1347

Van Cleve K, Chapin FS III, Dyrness CT & Viereck LA (1991) Element cycling in taiga forests: state-factor control. BioSci. 41: 78–88

VEMAP members (1995) Vegetation/ecosystem modeling and analysis project: Comparing biogeochemistry models in a continental-scale study of terrestrial ecosystem responses to climate change and CO_2 doubling. Glob. Biogeochem. Cyc. 9: 407–437

Vitousek PM & Howarth RW (1991) Nitrogen limitation on land and in the sea: How can it occur? Biogeochem. 13: 87–115

Vitousek PM, Aber JD, Howarth RW, Likens GE, Matson PA, Schindler DW, Schlesinger WH & Tilman D (1997a) Human alteration of the global nitrogen cycle: sources and consequences. Ecol. Applic. 7: 737–750

Vitousek PM, Chadwick OA, Crews T, Fownes J, Hendricks D & Herbert D (1997b) Soil and ecosystem development across the Hawaiian Islands. GSA Today 7(9): 1–8

Vitousek PM, Hedin LO, Matson PA, Fownes JH & Neff J (1998) Within-system element cycles, input-output budgets, and nutrient limitation. In: Pace M & Groffman P (Eds) Successes, Limitations, and Frontiers in Ecosystem Science (pp 432–452). Springer-Verlag, Berlin, Germany

Walker TW & Syers JK (1976) The fate of phosphorus during pedogenesis. Geoderma 15: 1–19

Biogeochemistry **46:** 203–231, 1999.

Do top-down and bottom-up controls interact to exclude nitrogen-fixing cyanobacteria from the plankton of estuaries? An exploration with a simulation model

ROBERT W. HOWARTH, FRANCIS CHAN & ROXANNE MARINO
Program in Biogeochemistry & Environmental Change, Corson Hall, Cornell University, Ithaca, NY 14853, U.S.A. (E-mail: rwh2@cornell.edu)

Received 10 December 1998

Key words: Baltic Sea, cyanobacteria, estuaries, grazing, iron, lakes, molybdenum, nitrogen, nitrogen fixation, nitrogen limitation, zooplankton

Abstract. Explaining the nearly ubiquitous absence of nitrogen fixation by planktonic organisms in strongly nitrogen-limited estuaries presents a major challenge to aquatic ecologists. In freshwater lakes of moderate productivity, nitrogen limitation is seldom maintained for long since heterocystic, nitrogen-fixing cyanobacteria bloom, fix nitrogen, and alleviate the nitrogen limitation. In marked contrast to lakes, this behavior occurs in only a few estuaries worldwide. Primary production is limited by nitrogen in most temperate estuaries, yet no measurable planktonic nitrogen fixation occurs. In this paper, we present the hypothesis that the absence of planktonic nitrogen fixers from most estuaries is due to an interaction of bottom-up and top-down controls. The availability of Mo, a trace metal required for nitrogen fixation, is lower in estuaries than in freshwater lakes. This is not an absolute physiological constraint against the occurrence of nitrogen-fixing organisms, but the lower Mo availability may slow the growth rate of these organisms. The slower growth rate makes nitrogen-fixing cyanobacteria in estuaries more sensitive to mortality from grazing by zooplankton and benthic organisms.

We use a simple, mechanistically based simulation model to explore this hypothesis. The model correctly predicts the timing of the formation of heterocystic, cyanobacterial blooms in freshwater lakes and the magnitude of the rate of nitrogen fixation. The model also correctly predicts that high zooplankton biomasses in freshwaters can partially suppress blooms of nitrogen-fixing cyanobacteria, even in strongly nitrogen-limited lakes. Further, the model indicates that a relatively small and environmentally realistic decrease in Mo availability, such as that which may occur in seawater compared to freshwaters due to sulfate inhibition of Mo assimilation, can suppress blooms of heterocystic cyanobacteria and prevent planktonic nitrogen fixation. For example, the model predicts that at a zooplankton biomass of 0.2 mg l^{-1}, cyanobacteria will bloom and fix nitrogen in lakes but not in estuaries of full-strength seawater salinity because of the lower Mo availability. Thus, the model provides strong support for our hypothesis that bottom-up and top-down controls may interact to cause the absence of planktonic nitrogen fixation in most estuaries. The model also provides a basis for further exploration of this hypothesis in individual estuarine systems and correctly predicts that planktonic nitrogen fixation can occur in low salinity estuaries, such as the Baltic Sea, where Mo availability is greater than in higher salinity estuaries.

Introduction

Net primary production in many of the earth's ecosystems, both aquatic and terrestrial, is limited by nitrogen. A variety of mechanisms such as denitrification and mobility of nitrate can contribute to nitrogen limitation, yet the widespread prevalence of such limitation is somewhat surprising given the ability of many types of bacteria to fix atmospheric N_2 and thereby alleviate shortages of nitrogen (Vitousek & Howarth 1991; Vitousek & Field, this volume). To understand how nitrogen limitation can persist over time, one must be able to understand why nitrogen fixation is sometimes unable to make up the deficit of nitrogen relative to phosphorus and other elements.

Estuaries and coastal seas of the temperate zone comprise one type of ecosystem in which net primary production is often limited by nitrogen (Nixon & Pilson 1983; D'Elia et al. 1986; Howarth 1988; NRC 1993) and in which rates of nitrogen fixation tend to be very low (Horne 1977; Fogg 1987; Howarth et al. 1988a). This behavior is in marked contrast to many temperate lakes, where P limitation of production is common (Schindler 1977; NRC 1993) and where nitrogen fixation rates are often high (Horne & Goldman 1972; Horne 1977; Flett et al. 1980; Howarth et al. 1988a). In fact, nitrogen fixation in lakes is one mechanism which leads to phosphorus limitation of production; generally in mesotrophic and eutrophic temperate lakes, when the N:P ratio is low, heterocystic species of cyanobacteria bloom in the plankton and fix nitrogen. This can alleviate the shortage of nitrogen and result in long term limitation of production by P (Schindler 1977; Flett et al. 1980; Howarth et al. 1988b). However, heterocystic cyanobacteria are rare or absent from the plankton of most estuaries and coastal seas even when these ecosystems are strongly nitrogen limited, and rates of nitrogen fixation in the plankton of most estuaries are immeasurably low (Horne 1977; Doremus 1982; Fogg 1987; Howarth et al. 1988a; Paerl 1990; NRC 1993). Two major exceptions are the Baltic Sea (Lindahl & Wallstrom 1985; Wallstrom 1988, 1991; Wallstrom et al. 1992; Niemisto et al. 1989; Moisander et al. 1996; Lehtimaki et al. 1997) and the Harvey-Peel Inlet in southwestern Australia (Huber 1986; Lukatelich & McComb 1986).

Ecologists have long been intrigued by the contrast between abundant nitrogen fixation by planktonic cyanobacteria in lakes and the relative absence of this process in most estuaries and coastal seas. The dichotomy between lakes and estuaries is so striking that one is compelled to search for a single factor to explain the difference, and the literature contains numerous hypotheses where one factor such as turbulence or a low availability of Mo (an element required for nitrogen fixation) is invoked to explain the relative lack of nitrogen fixation in coastal ecosystems compared to lakes (Doremus 1982; Paerl 1985; Howarth & Cole 1985; Fogg 1987; Valiela 1991). Nonetheless, no one explanation has gained widespread acceptance, and evidence inconsistent with or weakening of many of the single-control hypotheses has accumulated (Marino et al. 1990; Cole et al. 1993; Howarth et al. 1993, 1995).

Two trace metals, molybdenum (Mo) and iron (Fe), are both required for nitrogen fixation, and both are likely to be less available in seawater than in freshwaters (Howarth et al. 1988b; Marino et al. 1990). For Mo, the low availability is the result of competitive inhibition of molybdate uptake by sulfate. Even though dissolved Mo is fairly abundant in seawater compared to other trace metals (Howarth et al. 1988b), the thermodynamically stable form of Mo in oxic waters is molybdate, an anion which is stereochemically similar to sulfate. Thus, sulfate can inhibit Mo uptake by cyanobacteria and other phytoplankton (Howarth & Cole 1985; Cole et al. 1993). Inhibition of Mo assimilation by sulfate has also been seen in physiological studies with a variety of other organisms (Stout & Meagher 1948; Elliot & Mortenson 1975; Huising & Matrone 1975; Cardin & Mason 1976). In 1985, we suggested that the availability of Mo in estuaries may be sufficiently low as to exclude nitrogen-fixing cyanobacteria (Howarth & Cole 1985). However, our subsequent work has shown that sulfate only partially inhibits Mo uptake at the sulfate and Mo concentrations which characterize seawater (Marino et al. 1990; Cole et al. 1993). This suggests that while Mo availability may contribute to the relative absence of heterocystic cyanobacteria in the plankton of estuaries, it is unlikely to be the sole cause (Marino et al. 1990). The role of Fe in regulating nitrogen-fixation by planktonic cyanobacteria in estuaries has received much less study than has Mo (Howarth et al. 1988b; Marino et al. 1990; Vitousek & Howarth 1991), although the role of Fe in limiting the growth of oceanic phytoplankton has received much recent study (see for example, Martin et al. 1994), and Fe availability has been suggested as a regulator of nitrogen fixation in oceanic waters (Michaels et al. 1996; Falkowski 1997). We suspect that as with Mo, a relatively low availability of Fe in estuaries compared to lakes (Howarth et al. 1988b; Marino et al. 1990) may slow the growth rate of cyanobacteria but is unlikely to act as an absolute constraint.

We have come to believe that the relative lack of nitrogen fixation by plankton in estuaries compared to lakes is not the result of a single factor, but rather is caused by an interaction of a bottom-up control by availabilities of essential trace metals and a top-down control by grazing. In this paper, we present the results of a simulation model which explores this interaction.

The hypothesis

Our hypothesis is that low availabilities of Mo and/or Fe in estuaries lead to slow growth rates of planktonic, nitrogen-fixing cyanobacteria. This leaves the cyanobacteria vulnerable to consumption by generalized grazing animals, leading to very low numbers and thus low rates of nitrogen fixation. Our hypothesis does not state that herbivores in estuaries specifically seek out filamentous, heterocystic cyanobacteria as food (which would be a poor strategy, given the virtual absence of such cyanobacteria). Rather, we suggest that estuarine ecosystems contain generalized grazers which are capable of feeding on fine cyanobacterial filaments and are likely to do so incidentally to their feeding on other particles when filaments are present.

Compared to other phytoplankton, heterocystic cyanobacteria species that typically dominate nitrogen fixation in lakes may be particularly vulnerable to grazing (Schaffner et al. 1994; Epp 1995). Nitrogen fixation in these cyanobacteria occurs only in specialized, nonphotosynthetic cells called heterocysts (Bothe 1982; Gallon 1992). The energy needs for nitrogen fixation in the heterocyst are supported by other photosynthetic cells, and generally, many photosynthetic cells are required in a filament to build and support nitrogen fixation in one heterocyst (Turpin et al. 1985; Rowell & Kerby 1991). Thus, considerable filament growth must occur before any nitrogen fixation can occur. It is likely that if a filament is fully or partially grazed before enough photosynthetic cells are produced, the cyanobacteria will not produce heterocysts or fix nitrogen.

The growth rate of the cyanobacteria is an important parameter in our overall hypothesis. In freshwater studies, zooplankton have been found to feed on relatively fine and short cyanobacterial filaments, while larger and longer filaments are less susceptible to grazing (Schaffner et al. 1994; Epp 1995). Thus, fast-growing cyanobacteria are of a vulnerable size for only a short period of time until they grow sufficiently to become larger filaments. However, if low Mo and/or Fe availabilities in estuaries slow the growth rate of cyanobacteria, filaments will remain small and more vulnerable to grazing mortality for a longer period of time. Hence, the biological availabilities of Mo and Fe are not absolute constraints on growth, but may be factors which lower the growth rate of cyanobacteria in estuaries relative to lakes and thereby increase the vulnerability of the cyanobacteria to grazing.

Calanoid copepods (particularly *Acartia* species), ctenophores, and a variety of bivalves (mussels, oysters) are examples of filter feeding animals common in estuaries which may feed on cyanobacterial filaments. In a series of mesocosm experiments using seawater of salinity greater than 30‰, we found that planktonic, nitrogen-fixing cyanobacteria (*Anabaena* sp.) grew and fixed nitrogen when zooplankton populations were kept low (by adding zooplanktivorous fish) and benthic filter feeders were absent. However, the nitrogen-fixing cyanobacteria were suppressed by the addition of the blue mussel (*Mytilis edulis*) or higher densities of zooplankton in the mesocosms (Chan et al., manuscript in preparation). In short-term grazing experiments, we further demonstrated that *Acartia* sp. were able to feed upon cyanobacterial filaments from the mesocosms (Chan et al., manuscript in preparation), and consumption of filamentous cyanobacteria by *Mytilis edulis* has been reported by others (Falconer et a. 1992). Nitrogen-fixing, heterocystic cyanobacteria have not been reported in the plankton of any estuary in North America (Howarth et al. 1988a; Howarth & Marino 1990), including Narragansett Bay, Rhode Island (Karentz and Smayda 1998) which was the source of water for our mesocosm experiment. In fact, at salinities greater than 12‰, planktonic heterocystic cyanobacteria have been reported only for two estuaries: one in southwestern Australia (Huber 1986; Lukatelich & McComb 1986) and one in Tasmania (Jones et al. 1994). Our experiment thus provides strong evidence that a top-down control is likely to be important in excluding these nitrogen fixers from the plankton of estuaries. This influence and the interaction with the growth rate of the nitrogen-fixing cyanobacteria as influence by the availability of the trace metal Mo is explored in the model presented here.

Model structure

We constructed our model using STELLA ("research" version 4.0). The structure of the model is illustrated graphically in Figure 1. We have kept the structure quite simple so as to ease interpretation of model behavior. While we have used realistic parameters, we have specifically chosen not to duplicate much of the complexity of nature. For instance, while it is well known that increasing concentrations of inorganic nitrogen can suppress nitrogen fixation (Howarth et al. 1988b) and that the build up of inorganic nitrogen from nitrogen fixation over a growing season can contribute to the decline of nitrogen-fixing cyanobacteria (Wallstrom 1991), our model has no such feedbacks. The model is designed only to study the factors involved with the initiation of a bloom of nitrogen-fixing cyanobacteria, and not the decline of these blooms. This is justified since the purpose of the model is to examine why nitrogen-fixing blooms do not generally occur in estuaries and not to predict the extent or duration of such blooms once they form. The

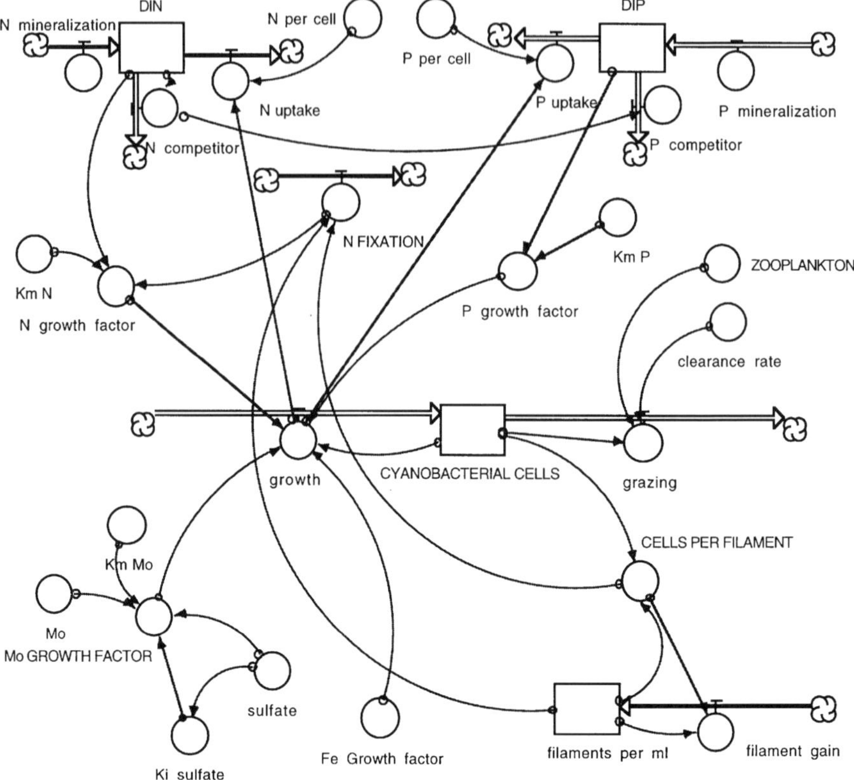

Figure 1. STELLA graphical representation of the simulation model. See text for equations underlying model structure.

initial conditions for the model are intended to be representative of temperate lakes immediately after the start of summer stratification. In the presentation below, we show the model as a series of differential equations. In practice, the STELLA software solves these as a series of difference equations using a time step of 0.1 hour.

The abundance of cyanobacterial cells (C) is calculated from their rate of growth (GROW) and their mortality from grazing by zooplankton (GRAZE).

$$dC/dt = (GROW - GRAZE) * C. \tag{1}$$

The initial condition is taken as 240 cyanobacterial cells per ml, a very low abundance. Our hypothesis is that grazing by both zooplankton and benthic filter feeders may be important in regulating blooms of heterocystic cyanobacteria. However, for the sake of simplicity, in the model we explicitly consider only grazing by crustacean zooplankton; grazing by benthic filter feeders or by other zooplankton (such as ctenophores) would have a similar effect. In the model, grazing is a function of zooplankton biomass and the rate of feeding by zooplankton (described as a clearance rate of volume of water filtered per zooplankton biomass per time).

$$GRAZE = Z * CLEAR, \tag{2}$$

where Z is the total biomass of crustacean zooplankton and CLEAR is the clearance rate. For any given model run, the zooplankton biomass is held constant. This is of course unrealistic, since zooplankton biomass varies over time in real ecosystems, and will be related both to the overall rate of primary production and to top-down controls from fish and other predators. However, holding zooplankton biomass constant makes comparisons among systems easier. For standard model runs, we set zooplankton biomass at 0.2 mg l^{-1}, a value frequently encountered in mesotrophic and eutrophic lakes and estuaries (Durbin & Durbin 1981; Pace 1986; Varmo et al. 1989; Tackx et al. 1990; Pace et al. 1992; White & Roman 1992; Uitto 1996; Canfield & Jones 1996; Elmgren & Larsson 1997). For clearance rate, we use a value of 5 ml hr^{-1} per mg of zooplankton. Studies for zooplankton feeding on cyanobacterial filaments have reported a range of clearance rates from 0.2 to over 200 ml hr^{-1} per mg of zooplankton, but the majority of these reported rates are greater than 5 ml hr^{-1} per mg of zooplankton (Holm et al. 1983; Haney 1987; Burns & Xu 1990; Schaffner et al. 1994; Burns & Hegarty 1994; Fulton 1988; James & Forsyth 1990; Sellner et al. 1994; Hanson et al. 1998). Thus, our chosen rate is relatively conservative. For any given model run the parameter GRAZE is a constant.

Cyanobacterial growth rate is a function of the intrinsic, maximum possible rate and several multipliers which lower the rate of growth based on availabilities of phosphorus, nitrogen, and molybdenum (referred to as "growth factors" in Figure 1).

$$GROW = 0.03 * P_{mult} * N_{mult} * Mo_{mult}. \tag{3}$$

The maximum growth rate of 0.03 hr^{-1} is reasonable for freshwater cyanobacteria (Gibson & Smith 1982) and is about twice the rate observed for planktonic cyanobacteria isolated from the Baltic Sea (Wallstrom 1991). The phosphorus-availability multiplier (P_{mult}) is simply a Michaelis-Menten function of dissolved inorganic phosphorus (DIP).

$$P_{mult} = DIP/(DIP + Km_P), \tag{4}$$

where Km_P is the half-saturation constant for DIP uptake. We assume a value of 2 μM for Km_P, a value typical for both freshwater and Baltic Sea nitrogen-fixing cyanobacteria (Nalewajko & Lean 1978; Wallstrom 1991). We handle the nitrogen-availability multiplier (N_{mult}) in an analogous manner for the case when no nitrogen fixation is occurring. However, when nitrogen fixation is occurring, we assume that this fully meets the nitrogen needs for growth and so nitrogen imposes no constraint on growth. This gives nitrogen-fixing cyanobacteria a competitive advantage when DIN concentrations are low.

$$\text{If N fixation} > 0,\ N_{mult} = 1, \tag{5}$$

$$\text{If N fixation} = 0,\ N_{mult} = \text{DIN} / (\text{DIN} + Km_N), \tag{6}$$

where Km_N is the half-saturation constant for the uptake of dissolved inorganic nitrogen (DIN). We assume a value of 20 μm for Km_N, a reasonably low value for filamentous cyanobacteria (Zevenboom & Mur 1978). For the molybdenum-availability multiplier (Mo_{mult}), we use an extended Michaelis-Menten type equation which also includes the inhibitory effect of sulfate (Cole et al. 1993):

$$Mo_{mult} = 4.2 * Mo/[Mo + Km_{Mo} * (1 + \text{sulfate}/Ki)], \tag{7}$$

where Km_{Mo} is the half-saturation "constant" for uptake of molybdate (in nM), Ki is the inhibition "constant" for the effect of sulfate on Mo uptake (in mM), Mo is the ambient molybdenum concentration in nM, sulfate is the ambient sulfate concentration in mM, and 4.2 is a scaling factor so that Mo_{mult} equals 1 for average freshwater concentrations of Mo and sulfate (5 nM and 0.11 mM, respectively; Marino et al. 1990). Cole et al. (1993) demonstrated that Km_{Mo} and Ki are not truly constants but rather vary in response to the concentrations of Mo and sulfate in natural waters. In waters where sulfate is high, the phytoplankton community responds to some extent so that per mole of sulfate, the sulfate is somewhat less inhibitory (larger Ki) than in systems where the sulfate concentration is lower. We estimate Ki as a function of sulfate concentration (in mM), following Cole et al. (1993):

$$Ki = 12 * \text{sulfate}/(0.3 + \text{sulfate}). \tag{8}$$

Thus, Ki is constant at 12 mM for ecosystems with high sulfate concentrations (greater than 10 to 15 mM). At progressively lower sulfate concentrations, Ki decreases (inhibition is proportionately greater per mole of sulfate, although the overall inhibition is still less since the molar concentration of sulfate is less).

From Mo assimilation data from several ecosystems as well as several cultures of cyanobacteria (benthic as well as planktonic), Cole et al. (1993) concluded that Km_{Mo} could be predicted as a function of the ambient Mo concentration (in nM) according to the following equation:

$$\log Km_{Mo} = 0.415 * \log Mo + 0.68. \tag{9}$$

However, this equation in combination with Equation 7 and Equation 8 suggests that Mo is actually more available in seawater (Mo = 110 nM; sulfate = 28 mM; Marino et al. 1990) than in average freshwaters, perhaps by more than 40%. This is strongly at variance with the observation that Mo is equally partitioned between particulate and dissolved phases in some freshwaters but is present overwhelmingly as the dissolved form in seawater; that is, independent evidence strongly shows that the biological availability of Mo is much less in seawater than in freshwater (Howarth et al. 1988b) since the ratio of dissolved to particulate metals is a good indicator of their biological availability in natural waters (Morel & Hudson 1985).

Therefore, we have further considered the analysis of Cole et al. (1993). Their analysis included data from cultures where the Mo concentration from the original source ecosystem is poorly known. Further, cyanobacteria may have evolved in the culture conditions; most commonly used culture media have concentrations of Mo and other trace metals which are much higher than seen in natural waters. During many generations of growth in Mo-rich media, the cyanobacteria may have increased their half-saturation values for Mo uptake. Gibson and Smith (1982) note that nutrient uptake systems in cyanobacteria both evolve and adapt physiologically to the ambient concentration of substrates. Using the data of Cole et al. (1993) for natural systems but deleting their culture data from the analysis yields the following equation to predict Km_{Mo} as a function of Mo concentration:

$$\log Km_{Mo} = 0.95 * \log Mo + 0.53. \tag{10}$$

Note that within the variation of the data of Cole et al. (1993), Equations 9 and 10 are not statistically distinguishable. Note also that Equation 10 has less scatter (r^2 = 0.74) than does Equation 9 (r^2 = 0.43). While using Equation 9 with Equations 7 and 8 suggests that Mo is more available in seawater than in average freshwaters, using Equation 10 in place of Equation 9 predicts that Mo availability is 58% less in seawater than in average freshwaters. This is more in line with other available data, such as the partitioning of dissolved and particulate Mo (Howarth et al. 1988b). Equation 10 also makes greater physiological sense: it predicts that Km_{Mo} values for any given system (average freshwater and seawater) are consistently somewhat greater than the Mo concentrations (by some 3-fold), whereas Equation 9 (the original equation from Cole et al. 1993) similarly predicts a Km_{Mo} value for freshwaters that is somewhat higher than the Mo concentration (by some 2-fold) but a Km_{Mo} value for seawater that is considerably lower than the Mo concentration (by 3-fold). It makes greater physiological sense that the half-saturation constant be more consistently related to the Mo concentration. We have therefore used Equation 10 to estimate Km_{Mo} values in our simulation model.

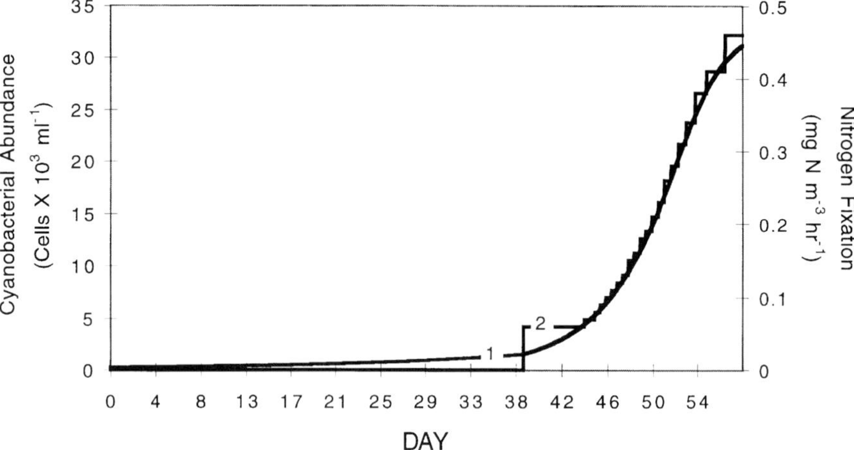

Figure 5. Model results for seawater concentrations of sulfate and Mo, as in Figure 4, except zooplankton biomass is reduced to zero from the standard model run conditions of 0.2 mg l^{-1}.

substantial error associated with estimating the extent to which sulfate inhibits Mo assimilation, so the actual change in growth rate may be more or less than we estimate. Nonetheless, the model clearly shows that relatively small decreases in cyanobacterial growth rate can have a profound effect on bloom initiation and on rates of nitrogen fixation. As we hypothesized, the ecological effect of zooplankton grazing can magnify the influence of a relatively small change in growth rate, so that the physiological effect of lower trace metal availability can result in the exclusion of heterocystic nitrogen fixers from the plankton. If Fe availability is less in estuaries than in lakes (as discussed further below), this too could have a similar effect.

The virtually complete suppression of nitrogen fixation over a two-month period in the seawater model run is not solely a result of the lower Mo availability, but rather results from the interaction of grazing and the slower growth rate caused by low Mo availability. Thus, when the model is run with seawater concentrations of sulfate and Mo but with the zooplankton biomass set at zero, a cyanobacterial bloom does eventually occur (Figure 5). The bloom occurs some 2 weeks later than in the standard freshwater run, but numbers of cyanobacterial cells and rates of nitrogen fixation are comparable to those in the standard freshwater model run, or perhaps even somewhat higher (Figure 5).

The Baltic Sea is one of the very few estuaries in the world where planktonic heterocystic cyanobacteria occur and fix nitrogen at reasonable rates. Our model correctly predicts this (Figure 6). The Baltic Sea is of relatively low salinity, with salinities ranging from near zero in the northern

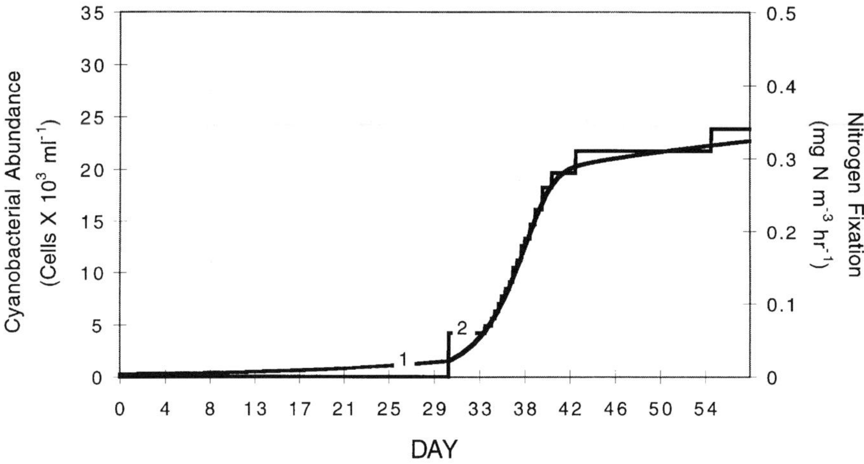

Figure 6. Model results for Baltic Sea concentrations of sulfate and Mo. All other parameters are as in the standard freshwater model run conditions as shown in Figure 2.

Gulf of Bothnia to 8 ‰ in the southern Baltic. The highest rates of nitrogen fixation are generally found in regions where the salinity is 6 ‰ or less (Lindahl & Wallstrom 1985; Wallstrom 1988, 1991; Wallstrom et al. 1992; Niemisto et al. 1989; Moisander et al. 1996; Lehtimaki et al. 1997). Since Mo and sulfate behave conservatively as a function of salinity in most oceanic waters (Howarth et al. 1988b) including the Baltic Sea (Howarth & Marino, unpublished data), we can estimate that at a salinity of 6 ‰, the sulfate concentration would be 4.8 mM and the Mo concentration would be 18 nM. Using Equations 8 and 10 to estimate the sulfate inhibition constant and Mo half saturation constant at these ambient concentrations, Equation 7 predicts a value for Mo_{mult} of 0.78. That is, the predicted rate of cyanobacterial growth is 22% less than for average freshwaters but is 86% faster than in full-salinity seawater (Howarth & Marino 1998). Figure 7 shows in general how we predict Mo availability will vary in estuaries as a function of salinity, expressed relative to average freshwater availability where the Mo_{mult} is taken as 1. Keeping the zooplankton biomass at 0.2 mg l^{-1} (the value used for the standard, freshwater run of the model, and a value typical for the coastal waters of the Baltic Sea; Varmo et al. 1989; Uitto 1996; Elmgren & Larsson 1997), our simulation model predicts that the initiation of a cyanobacterial bloom would be delayed by approximately 6 days in the Baltic Sea compared to average freshwaters, and that rates of nitrogen fixation would be virtually the same as in an average freshwater ecosystem after this short delay (compare Figure 6 with Figure 2).

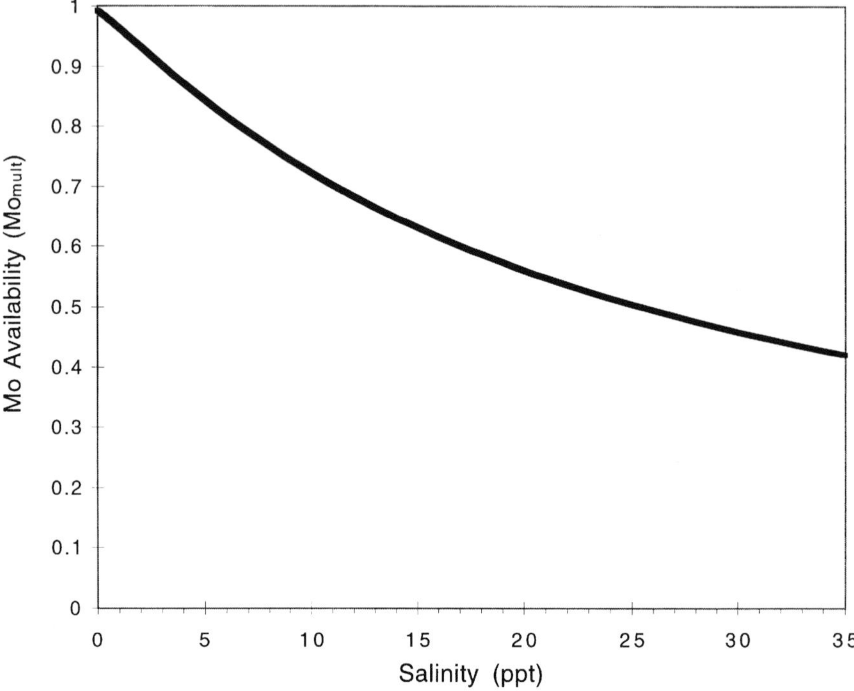

Figure 7. Estimated values of Mo availability in estuaries (expressed as the Mo_{mult} which is taken as a value of 1.0 for average freshwater values of Mo and sulfate) as a function of salinity; based on Equations 7, 8, and 10. A simple mixing model is used to estimate sulfate and Mo concentrations at any given salinity based on their concentrations in seawater and in average freshwaters. Mo availability decreases in a non-linear fashion as salinity increases from freshwater to full-strength seawater.

The only other estuaries in the world where heterocystic, nitrogen-fixing cyanobacteria have been reported among the plankton in waters of salinities greater than a few ‰ are in southwestern Australia and in Tasmania. Compared to most estuaries in North America or Europe, these estuaries undergo extreme salinity variations seasonally, with salinities varying from near zero to well over 30 ‰ (Hearn & Lukatelich 1990). Blooms of planktonic, filamentous cyanobacteria begin in these estuaries during the low-salinity events (Huber 1986; Lukatelich & McComb 1986; Jones et al. 1994). This is broadly consistent with our hypothesis: the low salinity events would increase cyanobacterial growth rates by increasing the availability of Mo and Fe, and might also decrease mortality from grazing by zooplankton and benthic animals if the animal species present at high salinities were lost by the sudden intrusion of low-salinity water.

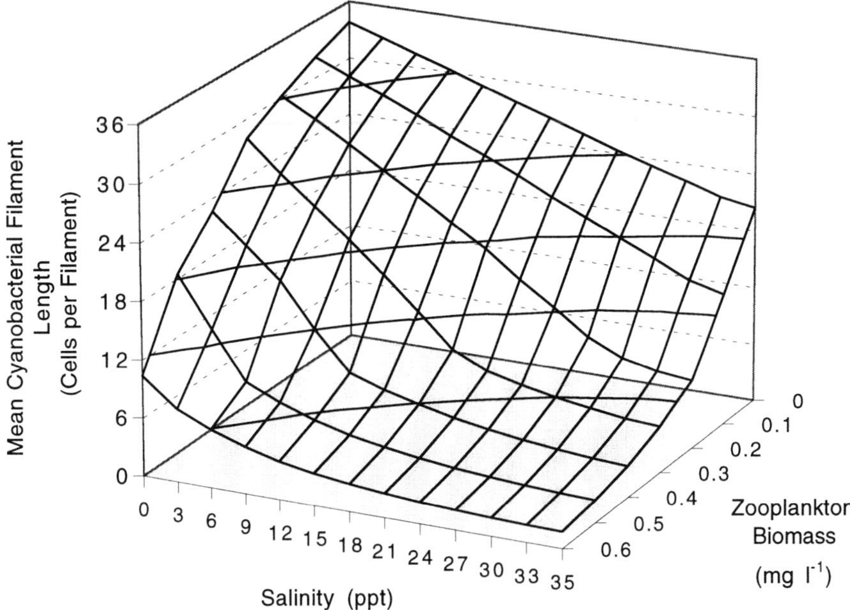

Figure 8. Model results illustrating responses of mean cyanobacterial filament length to variations in salinity and zooplankton biomass. Mean number of cells per filament (averaged over the course of a bloom) are depicted for individual model runs representing 91 combinations of salinity and zooplankton biomass. Salinity gradient ranges from "standard" freshwater conditions to full salinity seawater, while zooplankton biomass is set at 7 densities between 0.0 mg l^{-1} to 0.6 mg l^{-1}. Mo availability and growth factor are estimated based on Equations 7, 8 and 10.

Part of the explanation for the sensitivity of the model to both cyanobacterial growth rates and to grazing lies in the importance of having a sufficient number of photosynthetic cells in a cyanobacterial filament to support nitrogen fixation in the nonphotosynthetic heterocyst. The cyanobacteria have no competitive advantage over other phytoplankton until they begin to fix nitrogen, and they may in fact often be relatively poor competitors for assimilating DIN (as our model assumes). However, the cyanobacteria cannot fix nitrogen until they grow sufficiently long filaments, and this is difficult to do under significant grazing pressure (Figure 8). As salinity increases, the accompanying decrease in the availability of Mo accentuates the effects of zooplankton grazing on mean filament length (averaged over the entire time span of the model). Increasing zooplankton biomass from 0 to 0.2 mg ml^{-1} results in a rapid decline in mean cyanobacterial filament length from 20 to 7 cells per filament (Figure 8). For cyanobacteria growing in freshwater conditions, mean filament length declines only from 32 to 30 cells per filament over the same zooplankton biomass gradient (Figure 8).

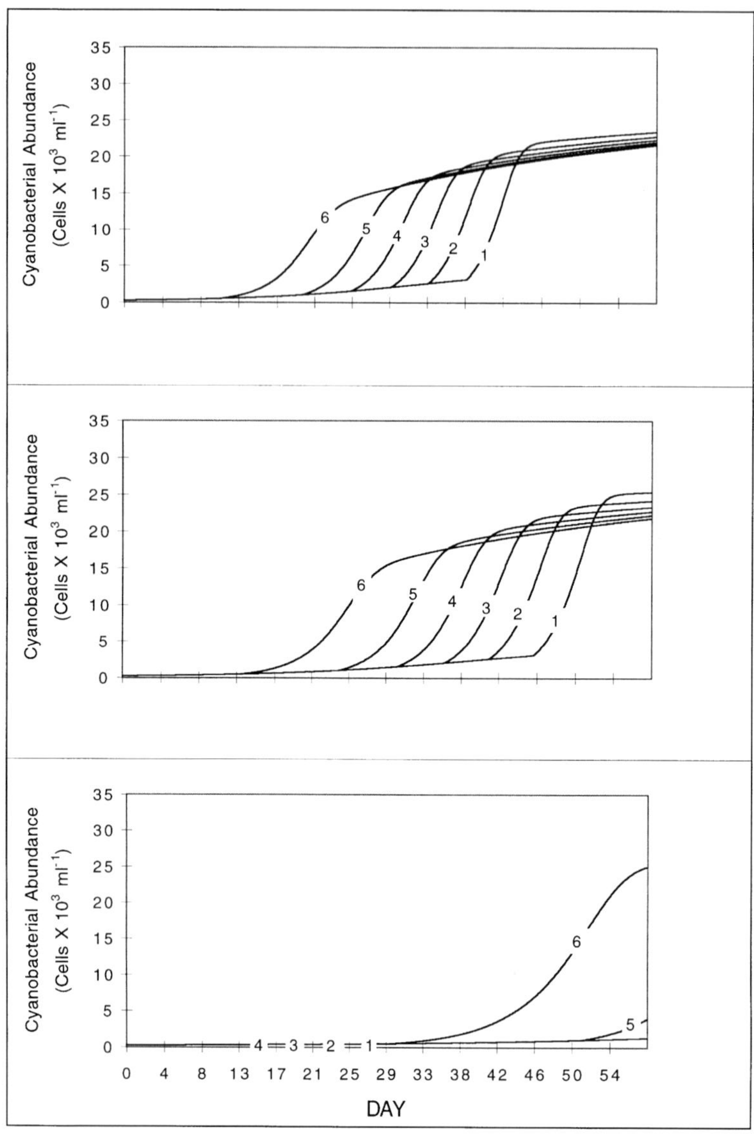

Figure 9. Sensitivity analysis showing model results for number of cyanobacterial cells over time as the number of photosynthetic cells assumed necessary to support the energetic needs of a heterocyst is altered. Top panel (9a) is for standard freshwater conditions, middle panel (9b) is for Baltic Sea concentrations of sulfate and Mo, and bottom panel (9c) is for seawater concentrations of sulfate and Mo. Other parameters are as in the standard model run shown in Figure 2. Line labeled "1" represents 38.5 cells per heterocyst, that labeled "2" represents 32 cells per heterocyst, that labeled "3" represents 25.5 cells per heterocyst, that labeled "4" represents 19 cells per heterocyst (the standard model run condition), that labeled "5" represents 12.5 cells per heterocyst, and that labeled "6" represents 6 cells per heterocyst. Values as low as 6 are outside of the range usually seen in nature.

For cyanobacterial filaments in natural systems, the number of photosynthetic cells associated with each heterocyst can vary widely from 12 to over 50 cells (Ogawa & Carr 1969; Horne & Goldman 1972; Horne et al. 1972; Kling et al. 1994; Gronlund et al. 1996). In the standard model run, we assume that 19 cells are needed to support the energetic needs of a heterocyst. In model runs where this number is increased (to 25.5 and to 32) or decreased (to 6 and to 12.5), the timing of bloom initiation is altered for conditions otherwise representative of the standard baseline freshwater run, but the eventual size of the bloom remains unchanged (Figure 9(a)). When we run this sensitivity analysis for Baltic Sea conditions of sulfate and Mo (and the standard zooplankton biomass value of 0.2 mg l^{-1}), the result is similar with blooms occurring approximately 6 days later than in the freshwater runs (Figure 9(b)). However, for seawater values of sulfate and Mo, the model shows no response except at the lowest number of cells per heterocyst (Figure 9(c)). Cyanobacterial cells begin to bloom in the seawater run within the 2-month time frame of the model only when we assume that 6 photosynthetic cells are sufficient to support a heterocyst. Such a low ratio of photosynthetic cells to heterocysts has not been observed in natural waters, probably because more cells are needed to support the energetic needs of heterocysts. Thus, while the model results show some sensitivity to the assumed number of cells needed to support a heterocyst, the overall comparison of freshwater and seawater results appears quite robust.

Presumably, grazing would place less of a constraint on the growth of unicellular, nonfilamentous cyanobacteria capable of nitrogen fixation than on heterocystic species. However, in lakes and in the few estuaries where planktonic nitrogen fixation has been observed, the organisms responsible for fixation are filamentous cyanobacteria and not unicellular organisms (Horne & Goldman 1972; Horne 1977; Flett et al. 1980; Howarth et al. 1988 a, b; Lindahl & Wallstrom 1985; Lukatelich & McComb 1986; Howarth & Marino 1998). Picoplankton-sized unicellular cyanobacteria (*Synechococcus* sp.) and heterotrophic bacteria can fix nitrogen, but in laboratory studies often require anoxic conditions to do so and have never been observed to fix nitrogen in oxic water columns, perhaps because of an inability to protect the nitrogenase enzyme from poisoning by oxygen (Gallon & Stal 1992; Bergman et al. 1997; Howarth & Marino 1998). The major nitrogen-fixing organism in oligotrophic oceanic waters, *Trichodesmium*, is colonial but nonheterocystic, with nitrogen fixation and photosynthesis occurring in the same cells (Carpenter & Capone 1992). We would therefore hypothesize that it may be less sensitive to the effects of grazing than are heterocystic cyanobacteria, such as those that are dominant nitrogen-fixers in lakes. This may in fact be part of the reason that this organism thrives in so much of the oligotrophic tropical and subtropical ocean water of the world; it may be constrained in growth rate by low availabilities of Mo and Fe, but a slow growth rate is tolerable without the need to produce long filaments before commencing nitrogen fixation. However, *Trichodesmium* is found in oligotrophic ocean waters, and has not been observed in the mesotrophic or eutrophic waters more characteristic of estuaries. The reasons for this remain unknown, although perhaps this species simply cannot compete well in higher nutrient environments (Howarth & Marino 1998).

Future model refinements

We have deliberately kept this version of our model simple so that interpretation is relatively straight forward. We have also only attempted to model the factors which regulate the initiation of cyanobacterial blooms and the maximum rates of nitrogen fixation, and not the factors which determine the length of bloom of the timing of bloom collapses. Even within this context, though, we see several ways in which the model could be refined in the future to further explore the controls on bloom initiation.

One refinement would be to include the inhibitory effect of DIN on heterocyst production and nitrogen fixation (Howarth et al. 1988b). The availability of some DIN is essential to support growth of cyanobacterial filaments before they become large enough to support a heterocyst and fix nitrogen, and the model currently includes this dynamic. However, DIN concentrations as low as 0.14 to 1.6 μM have been suggested to suppress heterocyst formation in a eutrophic lake (Horne et al. 1979), presumably because the cyanobacteria used the available DIN rather than pay the higher energetic cost of fixing nitrogen.

Another refinement would be to explicitly include the effect of benthic filter-feeding animals on cyanobacteria and on nitrogen fixation. From our mesocosm experiments (Marino et al., manuscript in preparation; Chan et al., manuscript in preparation), we know that benthic estuarine filter feeders such as blue mussels can feed on planktonic cyanobacteria, and their role as grazers may be as great as or greater than that of zooplankton in preventing the initiation of cyanobacterial blooms in estuaries. Freshwater filter feeders such as zebra mussels can also feed on planktonic cyanobacteria and may be important in regulating nitrogen fixation in some lakes.

A refinement to the zooplankton-grazing portion of the model would be to model the clearance rate at which zooplankton graze as a function of the abundance of cyanobacterial filaments. Several studies have found that at increasing densities of filaments, the clearance rate slows down, and some studies have even found a cessation of feeding at moderate to high densities of cyanobacterial filaments (Holm et al. 1983; Gliwicz & Lampert 1990; Schaffner et al. 1994). Our simulation model currently uses a constant clearance rate. Modifying the model so as to reduce clearance rates at higher filament densities would magnify the difference between the standard freshwater run (Figure 2) and the run with seawater concentrations of sulfate and Mo (Figure 4).

In our current model, zooplankton grazing removes cyanobacteria cells equally from all filaments, thereby shortening filaments (Equation 2). This is consistent with the behavior observed by Schaffner et al. (1994) for feeding by cladocerans on cyanobacteria. Our preliminary experiments with estuarine copepods (*Acartia* sp.) also suggest that they may feed on cyanobacterial filaments in this manner (Chan et al., manuscript in preparation). However, Schaffner et al. (1994) observed that freshwater copepods feed on cyanobacterial filaments by snipping pieces out of the middle, thereby not only eating some cells but creating much shorter filaments (and a greater number of filaments). We suspect that this type of feeding behavior would have an even greater influence on suppressing cyanobacterial blooms because of the large reduction in the number of photosynthetic cells in a filament; fewer cells are available to support the energetic requirements of the nonphotosynthetic heterocyst. The model could be used to explore the quantitative effect on nitrogen fixation of these contrasting feeding styles.

The model is parameterized to represent a mesotrophic or eutrophic ecosystem, and it contains no structural connection between zooplankton biomass and nutrient availability. In reality, ecosystems with greater nutrient availability have higher rates of primary productivity and support larger zooplankton populations (Pace 1986; Leibold et al. 1997). Such a linkage could be added to the model, perhaps simply relating zooplankton biomass to nutrient availability using regressions from comparative analyses. With such a change, the model could better explore the factors regulating nitrogen fixation in oligotrophic environments, including fixation by *Trichodesmium* in oceanic environments.

This version of our model does not consider the effect of Fe availability on cyanobacterial growth rate since we are unaware of adequate quantitative data on the relative availability of Fe in estuaries in comparison to lakes and on the effects of Fe availability on heterocystic cyanobacterial growth. Although Figure 1 suggests that there is an Fe-availability multiplier in the model, we have simply set this factor at 1. However, concentrations of both total Fe and dissolved Fe tend to be lower in estuaries than in lakes, so it is quite likely that the biological availability of Fe is also lower (Howarth et al. 1988b; Marino et al. 1990). Given the high Fe requirement of nitrogen-fixers and the low solubility of Fe in seawater, it seems quite plausible that

Fe may limit N-fixing activity in estuarine and marine waters (Rueter 1982; Howarth et al. 1988b; Brand 1991; Vitousek & Howarth 1991; Michaels et al. 1996; Falkowski 1997). Thus, our model probably underestimates the extent to which cyanobacterial growth rates in estuaries would be lower than in lakes.

Conclusions

We hypothesize that a low availability of Mo in estuaries compared to lakes does not act as an absolute physiological constraint against the occurrence of nitrogen-fixing cyanobacteria, but that low Mo availability may slow the growth rate of the cyanobacteria and thereby make them more sensitive to mortality from generalized grazing by zooplankton and benthic organisms. Our simple, mechanistically based simulation model indicates that this hypothesis is quite reasonable. Relatively small and environmentally realistic changes in either zooplankton biomass or Mo availability can result in the earlier or later initiation of cyanobacterial blooms, development of heterocysts, and subsequent nitrogen fixation. The model correctly predicts that increasing zooplankton biomass from 0.2 mg l^{-1} to 0.6 mg l^{-1} in freshwater ecosystems can suppress blooms of nitrogen-fixing cyanobacteria. The model also predicts that at a zooplankton biomass of 0.2 mg l^{-1}, cyanobacteria will bloom and fix nitrogen in lakes but not in estuaries of full-strength seawater salinity. Further, the model correctly predicts that planktonic nitrogen fixation can occur in low salinity estuaries, such as the Baltic Sea, and the model is consistent with the finding of nitrogen fixation in two estuaries in southwestern Australia and Tasmania as these experience drastic seasonal reductions in salinity.

Acknowledgements

This work was supported by grant DEB-9527405 from the Ecosystem Studies Program of the U.S. National Science Foundation. We thank Jon Cole for helpful discussion on Mo availability and Michael Pace, William Schaffner, and Nelson Hairston, jr. for their insights on zooplankton grazing. We also thank Michael Pace, Barbara Robson, and three anonymous reviewers for their comments on an early version of this manuscript.

References

Bergman B, Gallon JR, Rai AN & Stal LJ (1997) N_2 fixation by non-heterocystous cyanobacteria. FEMS Microbiol. Rev. 19: 139–185

Bothe H (1982) Nitrogen fixation. In: Carr NG & Whitton BA (Eds) The Biology of Cyanobacteria (pp 87–104). University of California Press, Berkeley, U.S.A.

Brand LE (1991) Minimum iron requirements of marine phytoplankton and the implications for the biogeochemical control of new production. Limnol. Oceanogr. 36: 1756–1771

Burns CW & Xu Z (1990) Calanoid copepods feeding on algae and filamentous cyanobacteria: Rates of ingestion, defecation and effects on trichome length. J. Plank. Res. 12: 201–213

Burns CW & Hegarty B (1994) Diet selection by copepods in the presence of cyanobacteria. J. Plank Res. 16: 1671–690

Canfield TJ & Jones JR (1996) Zooplankton abundance, biomass, and size-distribution in selected midwestern waterbodies and relation with trophic state. J. Fresh. Ecol. 11: 171–181

Cardin CJ & Mason J (1976) Molybdate and tungstate transfer by rat ileum: competitive inhibition by sulphate. Biochim. Biophys. Acta 455: 937–946

Carpenter EJ & Capone DG (1992) Nitrogen fixation in *Trichodesmium* blooms. In: Carpenter EJ, Capone DG & Rueter JG (Eds) Marine Pelagic Cyanobacteria: *Trichodesmium* and other Diazotrophs (pp 211–218). Kluwer, Dordrecht, The Netherlands

Cole JJ, Lane JM, Marino R & Howarth RW (1993) Molybdenum assimilation by cyanobacteria and phytoplankton in freshwater and salt water. Limnol. Oceanogr. 38: 25–35

D'Elia CF, Sanders JG & Boynton WR (1986) Nutrient enrichment studies in a coastal plain estuary; phytoplankton growth in large-scale, continuous cultures. Can. J. Fish. Aquat. Sci. 43: 397–406

Doremus C (1982) Geochemical control of dinitrogen fixation in the open ocean. Biol. Oceanogr. 1: 429–435

Durbin AE & Durbin EG (1981) Standing stock and estimated production rates of phytoplankton and zooplankton in Narragansett Bay, Rhode Island. Estuaries 4: 24–41

Elliott BB & Mortenson LE (1975) Regulation of molybdate transport by *Clostridium pasteurianum*. J. Bacteriol. 127: 770–779

Elmgren R & Larsson U (1997) Himmerfjarden: Forandringar i ett naringsbelastat kustekyosystem i Ostersjon. Rapport 4565. Naturvardsverket Forlag

Epp GT (1995) Herbivory in the freshwater plankton: interactions of *Daphnia pulicaria* and filamentous cyanobacteria. PhD dissertation, Cornell University, NY, U.S.A.

Falconer IR, Choice A & Hosja W (1992) Toxicity of edible mussels *Mytilus edulis* growing naturally in an estuary during a water bloom of the blue-green algae *Nodularia spumigena*. Environ. Toxicol. Water. Qual. 7: 119–123

Falkowski PG (1997) Evolution of the nitogen cycle and its influence on the biological sequestration of CO_2 in the ocean. Nature 387: 272–275

Flett RJ, Schindler DW, Hamilton RD & Campbell NER (1980) Nitrogen fixation in Canadian precambrian shield lakes. Can. J. Fish. Aquat. Sci .37: 494–505

Fogg GE (1987) Marine planktonic cyanobacteria. In: Fay P & Baalen CV (Eds) The Cyanobacteria (pp 393–414). Elsevier, Amsterdam, The Netherlands

Fulton RS (1988) Grazing on filamentous algae by herbivorous zooplankton. Fresh. Biol. 20: 263–271

Gallon JR (1992) Reconciling the incompatible: N_2 fixation and O_2. New Phytol 122: 571–609

Gallon JR & Stal LJ (1992) N_2 fixation in non-heterocystous cyanobacteria: An overview. In: Carpenter EJ, Capone DG & Rueter JG (Eds) Marine Pelagic Cyanobacteria: *Trichodesmium* and other Diazotrophs (pp 115–140). Kluwer, Dordrecht, The Netherlands

Gibson CE & Smith RV (1982) Freshwater plankton. In: Carr NG & Whitton BA (Eds) The Biology of Cyanobacteria (pp 463–489). Univ. of California Press, Berkeley, U.S.A.

Gliwicz ZM & Lampert W (1990) Food thresholds in *Daphnia* species in the absence and presence of blue-green filaments. Ecology 71: 691–702

Grobbelaar JU and House WA (1995) Phosphorus as a limiting resource in inland waters: interactions with nitrogen. In: Tiessen H (Ed.) Phosphorus in the Global Environment (pp 255–273). Wiley, Chichester

Gronlund L, Kononen K, Lahdes E and Makela K (1996) Community development and modes of phosphorus utilization in a late summer ecosystem in the central Gulf of Finland, the Baltic Sea. Hydrobiol. 331: 97–108

Haney JF (1987) Field studies on zooplankton-cyanobacteria interactions. New Zeal. J. Mar. Fresh. Res. 21: 467–475

Hansson L-A, Bergman E & Cronberg G (1998) Size structure and succession in phytoplankton communities: the impact of interactions between herbivory and predation. Oikos 81: 337–345

Hearn CJ & Lukatelich RJ (1990) Dynamics of Peel-Harvey Estuary, Southwest Australia. In: Cheng, RT (Ed.) Residual Currents and Long-term Transport. Coastal and Estuarine Studies 38 (pp 431–450). Springer-Verlag, New York, U.S.A.

Holm NP, Ganf GG & Shapiro J (1983) Feeding and assimilation rates of *Daphnia pulex* fed *Aphanizomenon flos-aquae*. Limol. Oceanogr. 28: 677–687

Horne AJ (1977) Nitrogen fixation – a review of this phenomenon as a polluting process. Prog. Water Technol. 8: 359–372

Horne AJ & Goldman CR (1972) Nitrogen fixation in Clear Lake, California. I. Seasonal variation and the role of heterocysts. Limnol. Oceanogr. 17: 678–692

Horne AJ, Dillard JE, Fujita DK & Goldman CR (1972) Nitrogen fixation in Clear Lake, California. II. Synoptic studies on the autumn *Anabaena* bloom. Limnol. Oceanogr. 17: 693–703

Horne AJ, Sandusk JC & Carmiggelt CJW (1979) Nitrogen fixation in Clear Lake, California. 3. Repetitive synoptic sampling of the spring *Aphanizomenon* blooms. Limnol. Oceanogr. 24: 316–328

Howarth RW (1988) Nutrient limitation of net primary production in marine ecosystems. Ann. Rev. Ecol. Syst. 19: 89–110

Howarth RW & Cole JJ (1985) Molybdenum availability, nitrogen limitation, and phytoplankton growth in natural waters. Science 229: 653–655

Howarth RW, Marino R, Lane J & Cole JJ (1988a) Nitrogen fixation in freshwater, estuarine, and marine ecosystems. 1. Rates and importance. Limnol. Oceanogr. 33: 669–687

Howarth RW, Marino R & Cole JJ (1988b) Nitrogen fixation in freshwater, estuarine, and marine ecosystems. 2. Biogeochemical controls. Limnol. Oceanogr. 33: 688–701

Howarth RW & Marino R (1990) Nitrogen-fixing cyanobacteria in the plankton of lakes and estuaries: A reply to the comment by Smith. Limnol. Oceanogr. 35: 1859–1863

Howarth RW, Butler T, Lunde K, Swaney D & Chu CR (1993) Turbulence and planktonic nitrogen fixation: a mesocosm experiment. Limnol. Oceanogr. 38: 1696–1711

Howarth RW, Swaney D, Marino R, Butler TJ & Chu CR (1995) Turbulence does not prevent nitrogen fixation by plankton in estuaries and coastal seas (reply to the comment of Paerl et al.). Limnol. Oceanogr. 40: 639–643

Howarth RW & Marino R (1998) A mechanistic approach to understanding why so many estuaries and brackish waters are nitrogen limited. In: Effects of Nitrogen in the Aquatic Environment, KVA Report 1998: 1 (pp 117–136). Kungl. Vetenskapsakademien (Royal Swedish Academy of Sciences), Stockholm

Huber AL (1986) Nitrogen fixation by *Nodularia spumigena* Mertens (Cyanobacteriaceae). 1: Field Studies and the contribution of blooms to the nitrogen budget of the Peel-Harvey Estuary, Western Australia. Hydrobiologia 131: 193–203

Huising J & Matrone G (1975) Biological interactions of sulfate and molybdate. Environ. Health Perspect. 10: 265

James MR & Forsyth DJ (1990) Zooplankton-phytoplankton interactions in a eutrophic lake. J. Plank. Res. 12: 455–472

Jones GS, Blackburn SI & Parker NS (1994) A toxic bloom of *Nodularia spumigena* Mertens in Orielton Lagoon, Tasmania. Aust. J. Mar. Fresh. Res. 45: 787–800

Karentz D & Smayda TJ (1998) Temporal patterns and variations in phytoplankton community organization and abundance in Narragansett Bay during 1959–1980. J. Plank Res. 20: 145–168

Kling HJ, Findlay DL & Komárek J (1994) *Aphanizomenon schindleri* sp. nov.: a new nostocacean cyanoprokaryote from the Experimental Lakes Area, northwestern Ontario. Can. J. Fish. Aquat. Sci. 51: 2267–2273

Lehtimaki J, Moisander P, Sivonen K & Kononen K (1997) Growth, nitrogen fixation, and nodularin production by two Baltic Sea cyanobacteria. Appl. Environ. Microbiol. 63: 1647–1656

Leibold MA, Chase JM, Shurin JB & Downing AL (1997) Species turnover and the regulation of trophic structure. Ann. Rev. Ecol. Syst. 28: 467–494

Lindahl G & Wallstrom K (1985) Nitrogen fixation (acetylene reduction) in planktonic cyanobacteria in Oregrundsgrepen, SW Bothnian Sea. Arch. Hydrobiol. 104: 193–204

Lukatelich RJ & McComb AJ (1986) Nutrient levels and the development of diatom and blue green algal blooms in a shallow Australian estuary. J. Plank. Res. 8: 597–618

Lynch M & Shapiro J (1981) Predation, enrichment, and phytoplankton community structure. Limnol. Oceanogr. 26: 86–102

Marino R, Howarth RW, Shamess J & Prepas EE (1990) Molybdenum and sulfate as controls on the abundance of nitrogen-fixing cyanobacteria in saline lakes in Alberta. Limnol. Oceanogr. 35: 245–259

Martin, JH, Coale KM, Johnson, KS, Fitzwater, SE, Gordon RM, Tanner SJ, Hunter CN, Elrod VA, Nowicki JL, Coley TL, Barber RT, Lindly S, Watson AJ, van Scoy K, Law CS, Liddicoat MI, Ling R, Stanton T, Stockelt J, Collins C, Anderson A, Bidigare R, Ondruske M, Latasa M, Millero FJ, Lee K, Yao W, Zhang JZ, Friederich G, Sakamoto C, Chavez F, Buck K, Kolber Z, Green R, Falkowski P, Chisholm SW, Hoge F, Swift R, Yungel J, Turner S, Nightingale P, Hatton A, Liss P & Tindale NW (1994) Testing the iron hypothesis in ecosystems of the equatorical Pacific Ocean. Nature 371: 123–129

Michaels AF, Olson D, Sarmiento JL, Ammerman JW, Fanning K, Jahnke R, Knap AH, Lipschultz F & Prospero JM (1996) Inputs, losses and transformations of nitrogen and phosphorus in the pelagic North Atlantic Ocean. Biogeochemistry 35: 181–226

Moisander P, Lehtimaki J, Sivonen K & Kononen K (1996) Comparison of $^{15}N_2$ and acetylene reduction methods for the measurement of nitrogen fixation by Baltic Sea cyanobacteria. Phycologia 35: 140–146

Morel FMM & Hudson RJ (1985) The geobiological cycle of trace elements in aquatic systems: Redfield revisited. In: Stumm W (Ed.) Chemical Processes in Lakes (pp 251–282). Wiley, New York, U.S.A.

Nalewajio C & Lean DRS (1978) Phosphorus kinetics – algal growth relationships in batch cultures. Mitt int. Verein. theor. angew. Limnol. 21: 184–192

Niemisto L, Rinne I, Melsvasalo T, Niemei Å (1989) Blue-green algae and their nitrogen fixation in the Baltic Sea in 1980, 1982, and 1984. Meri 17: 1–59

Nixon SW & Pilson MEQ (1983) Nitrogen in estuarine and coastal marine ecosystems. In: Carpenter EJ & Capone DG (Eds) Nitrogen in the Marine Environment (pp 565–648). Academic, New York, U.S.A.

NRC (1993) Managing wastewater in coastal urban areas. National Research Council, Washington, DC, U.S.A.

Ogawa FE & Carr JF (1969) The influence of nitrogen on heterocyst production in blue-green algae. Limnol. Oceanogr. 14: 342–351

Pace ML (1986) An empirical analysis of zooplankton community size structure across lake trophic gradients. Limnol. Oceanogr. 31: 45–55

Pace ML, Findlay SEG & Lints D (1992) Zooplankton in advective environments: the Hudson River community and a comparative analysis. Can. J. Fish. Aquat. Sci. 49: 1060–1069

Paerl HW (1985) Microzone formation: Its role in the enhancement of aquatic N_2 fixation. Limnol. Oceanogr. 30: 1246–1252

Paerl HW (1990) Physiological ecology and regulation of N_2 fixation in natural waters. In: Marshall KC (Ed.) Advances in Microbial Ecology (pp 305–343). Plenum, NY, U.S.A.

Rowell A & Kerby NW (1991) Cyanobacteria and their symbionts. In: Dilworth MJ & Glenn AR (Eds) Biology and Biochemistry of Nitrogen Fixation. Studies in Plant Science 1 (pp 373–407). Elsevier, New York, U.S.A.

Rueter JG (1982) Theoretical Fe limitations of microbial N_2 fixation in the oceans. EOS 63: 945

Schaffner WR, Hairston NG, jr & Howarth RW (1994) Feeding rates and filament clipping by crustacean zooplankton consuming cyanobacteria. Verh. Internat. Verein. Limnol. 25: 2375–2381

Schindler DW (1977) Evolution of phosphorus limitation in lakes. Science 195: 260–262

Sellner KG, Olson MM & Kononen K (1994) Copepod grazing in a summer cyanobacteria bloom in the Gulf of Finland. Hydrobiologia 292/293: 249–254

Stout PR & Meagher WR (1948) Studies of the molybdenum nutrition of plants with radioactive molybdenum. Science 108: 471–473

Tackx MLM, Bakker C & Van Rijswijk P (1990) Zooplankton grazing pressure in the Oosterschelde (The Netherlands). Neth. J. Sea Res. 25: 405–415

Turpin DH, Layzell DB & Elrifi IR (1985) Modeling the C economy of *Anabaena flos-aquae*. Plant. Physiol. 78: 746–752

Uitto A (1996) Summertime herbivory of coastal mesozooplankton and metazoan microplankton in the northern Baltic. Mar. Ecol. Prog. Ser. 132: 47–56

Valiela I (1991) Ecology of coastal ecosystems. In: Barnes RSK & Mann DH (Eds) Fundamentals of Aquatic Ecology (pp 57–76). Blackwell Scientific, Oxford

Varmo R, Viljamaa H, Pesonen L & Rinne I (1989) Two manipulated inner bays in the Helsinki Sea area, Northern Gulf of Finland. Aqua. Fenn. 19: 67–74

Vitousek PM & Field CB (this volume) Ecosystem constraints to symbiotic nitrogen fixers: A simple model and its implications. Biogeochemistry: in press

Vitousek PM & Howarth RW (1991) Nitrogen limitation on land and in the sea: how can it occur? Biogeochemistry 13: 87–115

Wallstrom K (1988) The occurrence of *Aphanizomenon flos-aquae* (Cyanophyceae) in a nutrient gradient in the Baltic. Kieler Meeresforschungen Sonderheft 6: 210–220

Wallstrom K (1991) Ecological studies on nitrogen fixing blue-green algae and on nutrient limitation of phytoplankton in the Baltic Sea. PhD thesis, Uppsala University. Uppsala, Sweden

Wallstrom K, Johansson S & Larsson U (1992) Effects of nutrient enrichment on planktonic blue-green algae in the Baltic Sea. Acta. Phytogeogr. Suec. 78: 25–31

White JR & Roman MR (1992) Seasonal study of grazing by metazoan zooplankton in the mesohaline Chesapeake Bay. Mar. Ecol. Prog. Ser. 86: 251–261

Zevenboom W & Mur LR (1978) Nitrogen uptake and pigmentation of N-limited chemostat cultures and natural populations of *Oscillatoria agardhii*. Mitt int. Verein. theor. angew. Limnol. 21: 261–274

Biogeochemistry **46:** 233–246, 1999.

The presence of nitrogen fixing legumes in terrestrial communities: Evolutionary vs ecological considerations

TIMOTHY E. CREWS
Environmental Studies Program, Prescott College, Prescott, Arizona 86301 U.S.A.

Received 10 December 1998

Key words: biogeography, Fabaceae, legumes, nitrogen fixation, nutrient limitation

Abstract. Nitrogen is often a limiting factor to net primary productivity (NPP) and other processes in terrestrial ecosystems. In most temperate freshwater ecosystems, when nitrogen becomes limiting to NPP, populations of N-fixing cyanobacteria experience a competitive advantage, and begin to grow and fix nitrogen until the next most limiting resource is encountered; typically phosphorus or light. Why is it that N-fixing plants do not generally function to overcome N limitation in terrestrial ecosystems in the same way that cyanobacteria function in aquatic ecosystems? To address this question in a particular ecosystem, one must first know whether the flora includes a potential set of nitrogen fixers. I suggest that the presence or absence of N-fixing plant symbioses is foremost an evolutionary consideration, determined to a large extent by constraints on the geographical radiation of woody members of the family Fabaceae. Ecological factors such as competition, nutrient deficiencies, grazing and fire are useful to explain the success of N-fixing plants only when considered against the geographical distribution of potential N-fixers.

Introduction

Plants are predominantly made up of carbon, oxygen and hydrogen which are supplied by air and water. Beyond these three elements, nitrogen is required in the greatest quantity. Although almost 80% of the atmosphere is comprised of nitrogen, which occurs in the gaseous form N_2, it cannot be directly accessed by plants.

The transformation, or 'fixation' of nitrogen from the unavailable gaseous form in the atmosphere to forms that plants and other organisms can use (either NH_4^+ or NO_3^-) is mediated by (1) bacteria in symbiotic relationships with vascular plants, (2) symbioses between cyanobacteria and fungi (lichens) or plants, (3) free living heterotrophic or autotrophic bacteria that are typically associated with soil or detritus, and (4) abiotic reactions in the atmosphere associated with lightening (Sprent & Sprent 1990). Although the latter three sources of fixation are more global in occurrence, the bacteria/vascular plant symbioses usually sustain the highest rates of nitrogen fixation per unit area where they occur (Boring et al. 1988). For the rest of the paper, vascular plants that have the capacity to host symbiotic relationships with N-fixing bacteria will be referred to as 'nitrogen fixing plants' or simply 'nitrogen fixers.' In addition, 'legumes' will be used interchangeably with members of the Fabaceae family.

Why nitrogen limitation commonly develops on land has perplexed ecologists for some time. Nitrogen is the sole nutrient whose rate of input is largely controlled by the biota. The input rates of all other essential nutrients can be influenced by the biota, but are ultimately constrained by abiotic factors that drive soil and rock dissolution (Gorham et al. 1979). Presumably, the organisms that are capable of fixing nitrogen should experience a competitive advantage when nitrogen is limiting, making N limitation a transient condition (Vitousek & Howarth 1991). Working to understand why N-fixers are present in some communities and stages of succession while not in others is important in answering more basic questions about the occurrence of nitrogen limitation in terrestrial ecosystems.

A number of ecological explanations have been suggested to help understand why nitrogen fixing plants are often *not* favored over non-fixing plants when N is limiting (Vitousek & Howarth 1991). Probably the most broadly applied explanation focuses on the relative energetic costs of supporting nitrogen fixation versus producing roots to appropriate soil nitrogen (Gutschick 1980, 1981; Vitousek & Field, this volume). Calculations and models suggest that it is energetically advantageous (i.e., requires less of a plant's photosynthate) to grow roots and take up soil N than to fix N when soil N is available (Gutschick 1981; Vitousek & Field, this volume). The implication of this energetic tradeoff is that when light becomes limiting during succession, N-fixing plants are outcompeted by non-fixing plants that are able to put relatively more energy into aboveground growth; thus nitrogen fixing plants are eliminated from the community before adequate N is fixed to attain maximum rates of NPP (Vitousek & Howarth 1991).

Ecological constraints that may limit the success of N-fixing plants other than the energetic cost of N fixation include the availability of soil nutrients other than N (especially P or Mo) (Robson & Bottomley 1991; Smith 1992; Crews 1993), the existence of poor edaphic conditions such as high acidity, alkalinity or aridity (Alexander 1984; Bordeleau & Prévost 1994), removal of N-fixing species by preferential grazing (Hulme 1994, 1996), and removal of woody dicots (including woody legumes) by fire (Bahre 1995).

I argue here that before ecological explanations for the success of N-fixing plants can be considered, it is essential to first evaluate the abundance or paucity of potential nitrogen fixing species – particularly woody legume species – for a given ecosystem. Some terrestrial ecosystems may be nitrogen limited simply because they are outside of the center of legume diversity. In ecosystems where nitrogen limitation exists even when potential fixers are in the flora, then ecological explanations for understanding the colonization and activity of N-fixers are appropriate and useful. Below I will review how the geographic radiation of the Fabaceae has resulted in the domination of the tropics by this group of plants and the comparative paucity of legumes in the floras of temperate regions.

Legume diversity and biogeography

The tropics

The family Fabaceae (Leguminosae) is by far the most diverse and widespread group of plants that have the capacity to host N-fixing bacteria. Ranking behind only the Asteraceae and Orchidaceae in size, the Fabaceae is the third largest family of flowering plants with about 650 genera and 18,000 species (Polhill et al. 1981; Sprent 1995). Although there is some debate as to how the Fabaceae should be broken up into sub-families, the predominant view is that there are three: the Caesalpinioideae, the Mimosoideae and the Papilioniodeae (Polhill et al. 1981; Sprent 1995). The Caesalpinioideae and Mimosoidae mainly consist of woody shrubs and trees (and woody vines in the case of Mimosoidae) that are largely confined to tropical and subtropical regions (NAS 1979). The Papilionoideae is made up of woody shrubs and trees as well as perennial and annual herbs; this sub-family is distributed worldwide with the woody members being largely concentrated in the tropics and sub-tropics (Allen & Allen 1981).

Only around 20% of all legume species and about half of legume genera have been examined for nodulation (Sprent 1994a; de Faria et al. 1989) and far less have been actually tested for nitrogen fixing activity (Virginia et al. 1989; Boring et al. 1988). Members of the three sub-families have been found to have different capacities to support nodulation. Of the species examined, 97% of the Papilionoideae, 90% of the Mimosoideae and 23% of the Caesalpinoideae have been found to nodulate (Allen & Allen 1981, de Faria et al. 1989). The N-fixing bacteria that infect the roots of Fabaceae belong to one of two genera, *Rhizobium* or *Bradyrhizobium* (Sprent 1994b).

It is generally accepted that legumes evolved in the humid tropics (Herendeen et al. 1992). They are well-represented in successional to mature, full-statured communities ranging from wet, lowland rainforests, to tropical deciduous forests, thorn scrub forests, deserts and savannas. Although it is

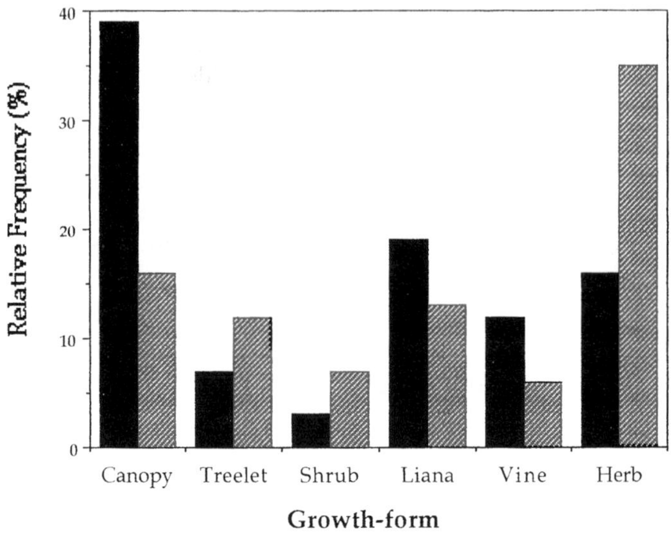

Figure 1. Relative frequency distribution of growth-forms for legumes (solid bars) and the total flora (hatched bars) on Barro Colorado Island, Panama. From Rundel (1989), data from Croat (1978).

clear that not all legumes can support nodulation and thus N fixation, the geographical extent of potential fixers is nevertheless ubiquitous in the tropics (Allen & Allen 1981; de Faria et al. 1989; Moreira de Souza et al. 1992).

Legumes comprise an important component of canopy trees in lowland mesic and wet tropical forests throughout the world (Rundel 1989). In lowland Neotropical rainforests, there are generally more legume tree species than any other family (Hammel 1990; Foster & Hubbell 1990; Foster 1990). Moreover, Moreira de Souza and colleagues (1992) report that in the Brazilian Amazon, the Fabaceae has the greatest diversity of all plant families. In measuring ecological success as a function of density, it is also clear that legumes are extremely successful in the wet tropics. Reports of legume densities range from 2–8% from the Africa and Asia (Rundel 1989). In the Neotropics, legumes commonly make up 12–15% of total tree stems (Rundel 1989) with reports exceeding 50% (Allen & Allen 1981). It should be re-emphasized that not only do legumes comprise a significant percentage of the floras and stand densities of lowland tropical forests, but that they are fully represented and in many cases dominant in the canopy, such as at Barro Colorado Island in Panama (Figure 1) (Croat 1978; Rundel 1989).

Legumes have also been extremely successful in drier tropical ecosystems. In dense stands of tropical deciduous forest at Chamela, Mexico, legumes are the most diverse family, making up about 14% of the overall flora (Martínez-

Yrízar et al. in press). In the drier tropical thorn-scrub ecosystems of Mexico, Gentry (1942) found legumes to be overwhelmingly dominant, with some communities having up to 90% coverage. Species of the legume genus *Acacia* are common to dominant in tropical thorn scrub communities in both Africa and Australia (see Rundel 1989).

Desert communities that evolved from tropical floristic elements, tend to have a significant legume presence. The Sonoran desert of the S.W. United States and N.W. Mexico is an example of a tropically-derived desert flora (Brown 1994), with a strong legume (herbaceous and woody) component (Turner et al. 1995; Eskew & Ting 1978). The two deserts adjacent to the Sonoran – the Chihuahuan and Mojave – are both derived from temperate floristic elements and, with the exception of *Acacia greggii* in the Chihuahuan, have considerably lower legume diversity and abundance (Rundel & Gibson 1996; Brown 1994).

Legumes are well represented in many, although not all, tropical savanna ecosystems. From the standpoint of diversity, Solbig (1996) reported that the Fabaceae has more common tree species in American savannas than any other plant family. In terms of abundance, Medina and Bilbao (1991) found that percent relative cover of legumes in Trachypogon-savanna sites in Venezuela ranged between 6–56%. Others have found low coverage by legumes despite the fact that on a per area basis, legume species richness was higher than that of grasses (Blydenstein 1967; Velasquez 1965 as cited in Medina & Bilbao 1991).

Overall, the rule in lowland American tropics and many tropical ecosystems on other continents, is that potentially nodulating, woody legumes are common and often dominant. This is true from open-canopied arid ecosystems to dense-canopied rainforests, from ecosystems with alkaline, base saturated soils to acid, P deficient soils.

In considering the evolutionary pathways that led to the incredible diversity and dominance of legumes in the tropics, McKey (1994) postulated that nitrogen fixing symbioses between legumes and rhizobia originally evolved in a nitrogen rich environment of mesic tropical forests. He offers evidence suggesting that in general, members of the Fabaceae are nitrogen-demanding plants that maintain high levels of N in leaf tissue in order to maximize photosynthetic rates per unit leaf area. This, in turn, allows the legumes to amortize the carbon costs of leaf construction, and thus have the ability to produce and drop leaves quickly in response to changing resource availability. McKey (1994) contends that this 'N-rich lifestyle' pre-dated the evolution of N-fixing symbioses; in fact, he suggests that it was specifically this N-demanding strategy that provided the selective pressure for the evolution of N-fixing symbioses. Once such symbioses in legumes evolved,

Interactions between N and water availability may differ for temporal versus geographic gradients of precipitation. Many researchers have seen greater responses to N addition in wet years than in dry years within a given location (Rogler & Lorenz 1957; Kilcher 1958; Smika et al. 1965; Owensby et al. 1970; Lorenz & Rogler 1972). Within a given location, however, plant production response to additional resources may be constrained by limitations of the extant plant community. Chapin et al. (1986) suggest that slow growing species in nutrient limited sites have relatively small responses to increases in resources above what they are accustomed to experiencing. Lauenroth and Sala (1992) and Burke et al. (1997) observed that, in nonfertilized sites, production responded less within a site to a given increase in annual precipitation resulting from yearly variation in rainfall than did production at different sites resulting from geographic variation in precipitation. If growth constraints exist, will there be a greater N response across a geographic precipitation gradient than across temporal variation in precipitation?

We wanted to test several hypotheses by using a literature review of fertilization experiments in arid, semi-arid, and subhumid regions:

(1) As ecosystem water availability increases, plant growth becomes more limited by N than by water. Across broad precipitation gradients, we would expect a greater relative response of production to N addition in more mesic locations than in drier locations, even if N availability also increases. A corollary to this would be that in experiments manipulating both water and N availability, NPP response to water addition would be greater than response to N addition at the drier end of the moisture gradient and vice versa at the wetter end.

(2) Water limits production more in dry years, N more in wet years. Therefore, within a given site, we would expect a greater relative response to fertilization in wetter years than in drier years (Figure 1).

(3) Production response to N addition will increase more as precipitation increases across geographical precipitation gradients than it will as precipitation increases within sites. (i.e., response 1 is greater than response 2, above). This is analogous to the response to precipitation for unfertilized conditions seen by others (Lauenroth & Sala 1992; Burke et al. 1997).

The second objective with our literature data set was to assess the utility of different indices of N limitation when comparing a wide variety of sites that differ substantially in a number of ecological characteristics (e.g., water availability, soil type, management regime). We used four indices that reflect different aspects of N limitation across precipitation gradients. First, systems that are more limited by N would be expected to gain more absolute biomass in response to fertilization than systems that are limited or co-limited by other resources. We refer to this absolute gain in biomass in response to fertilization as the *Absolute Increase* index of N limitation. Second, more N limited systems might have greater relative gains in biomass (i.e., the percent increase above control level) than less N limited systems. For example, a system that starts at 50 $g \cdot m^{-2} \cdot yr^{-1}$ of production may not add as much absolute biomass as a system that starts at 400, even if it is more N limited; however, the proportional increase in biomass may be greater. We refer to this relative change in production in response to fertilization as the *Relative Response* index of N limitation. Third, more N limited systems could have greater gains in plant production than less N-limited systems for the same amount of added N. We refer to the biomass gained per unit N added as the *Fertilizer Use Efficiency* (FUE) index of N limitation (to avoid confusion with other definitions of N Use Efficiency). This index is probably most useful at intermediate levels of fertilization: if N addition levels are low, FUE may be influenced by differences in microbial immobilization; if N addition levels are high, FUE will be influenced by saturation of the plant response. Fourth, Chapin et al. (1986) recommended using high levels of fertilization to saturate microbial immobilization to best assess nutrient limitation in plant communities. Therefore, more N limited systems might have a greater absolute response of production than less N limited systems at levels of N addition that are sufficiently high to saturate both microbial and plant demand. We refer to this maximum production response at high levels of added N as the *Maximum Response* index of N limitation. We used these four indices to investigate whether N limitation varies with water availability in dryland ecosystems, and to try to better understand some of the mechanisms underlying such variation.

Methods/data collection

Overview

We did a literature search for experiments on plant production response to N fertilization in arid, semi-arid, and subhumid ecosystems (mean annual precipitation = 200–1100 mm). Our indicator of N limitation was response to fertilization of aboveground plant growth. As independent moisture variables, we used mean annual precipitation (MAP) across the geographic gradient and actual annual precipitation (AAP) for temporal gradients within sites. The advantages of this approach are that we were able to gather large amounts of data from many different ecosystem types. This allowed us to look for broad scale patterns of N limitation using real data. A weakness is that patterns may be obscured by other variables such as differences in climate (e.g., mean annual temperature), topoedaphic features (slope, aspect, soil type and soil depth – which affect water availability to plants), management regime, and plant species composition. Wherever possible, we recorded information about such ecosystem characteristics in an attempt to account for these differences.

While annual precipitation is an imperfect index of plant water availability, it is reported in most experiments and a number of previous studies have found a strong correlation between MAP and plant production in dryland ecosystems (Noy-Meir 1973; Le Houerou & Hoste 1977; Le Houerou et al. 1988; Sala et al. 1988; Epstein et al. 1996). A truer index of plant water availability would incorporate actual evapotranspiration (AET), topography (runon and runoff, slope and aspect), and soil characteristics (e.g., depth and water holding capacity). Similarly, differences in the intensity and distribution of precipitation also influence plant water availability. However, these data are not available for most studies. We therefore used MAP and AAP with the assumption that across sites and across years, they would provide reasonable indices of water availability. Where the data allowed, we investigated effects of differences in site mean annual temperature (MAT) on responses to N because of potentially important temperature effects on plant water availability. For example, 313 mm of precipitation in Manyberries, Alberta (MAT = 4.6 °C) can mean substantially more plant available water than 315 mm of precipitation in Maipu, Chile (MAT = 14.5 °C). Because most of the studies surveyed provided no information on site temperature regime, we obtained MAT normals for 1966–1996 from the National Climate Data Center web site for locations in the United States (http://www.ncdc.noaa.gov) and for 1960–1990 from the Canadian Meteorological Centre (http://www.cmc.ec.gc.ca/climate/) for sites in Canada. Where exact location matches were not available, and for sites outside North America, we used information from the nearest available site or city.

We restricted the scope of our literature search by several criteria. We did not use studies of row crops and we used only fertilization data for N addition alone. A number of studies (e.g., at Cedar Creek LTER in Minnesota; Tilman 1987) also added phosphorus or other nutrients to all plots in a non-factorial manner, so we did not include these in our analyses. Where other nutrients were added in factorial, we used only control and +N data unless there was clearly no effect of the other treatments. We also restricted our analyses to studies in which N additions occurred in the year of production measurements. A number of other studies investigated residual effects of N addition – that is, the plant production response across several years to a single addition of N. In these cases, we only used data for the first year of N addition. We also did separate analyses for a subset of the experiments which included factorial N and water additions.

Data analysis

N limitation across geographic gradients of precipitation

We used data for aboveground plant biomass and aboveground net primary production (ANPP) of the entire ecosystem, not of individual species (except for two studies from desert ecosystems (Ettershank et al. 1978; Stephens & Whitford 1993)). While knowing the response of belowground production would certainly be helpful, these data are exceptionally sparse. Our indices of N limitation were calculated as follows:

Absolute Increase = slope of regression of ANPP vs fertilizer N; regressions were across many sites or experiments, but within levels of precipitation (see below);

Relative Response = (fertilized ANPP – control ANPP)∗100/control ANPP;

Fertilizer Use Efficiency as FUE = (fertilized ANPP – control ANPP)/ annual fertilizer N addition, with FUE in g of production/g added N;

Maximum Response to fertilization = (fertilized ANPP – control ANPP) at the highest levels of fertilization encountered in the studies surveyed (20–100 g $N \cdot m^{-2} \cdot yr^{-1}$).

For analyses involving the geographic precipitation gradient where studies had multiple years of data, we averaged the yearly data to get one overall estimate of response to N addition. We found 40 studies covering 42 different locations ranging in MAP from 211 mm to 1031 mm (control n = 98 data points, fertilized n = 157 data points). Levels of fertilization ranged from less than 1 to greater than 100 g $N \cdot m^{-2} \cdot yr^{-1}$.

We analyzed the geographic data set in two different ways. In the first set of analyses (for the Absolute Increase index), we categorized sites into 5 levels of MAP, regressed plant production against level of N fertilization within each of these precipitation levels, then compared the slopes and intercepts of the regressions for sites at different levels of MAP. The MAP levels were 200–300 mm, >300–450 mm, >450–600 mm, and >750–1100 mm. We refer to these as MAP levels 300, 450, 600, and 900, respectively; we had no data in the 600–750 mm level. We restricted the analysis to levels of fertilization < 25 g $N \cdot m^{-2} \cdot yr^{-1}$ to avoid complications of saturation of production at high levels of fertilization.

To assess the effects of other site variables, we divided plant composition into 3 categories depending on whether natives, exotics or both dominated communities in the experiments. We categorized sites as to whether they were grazed during the experiment, were grazed up until the year before the experiment but not during, were grazed within five years previous to the experiment but not during, or were not grazed within a least five years of the experiment. These categories were chosen in hopes of reflecting potential grazing effects on litter accumulation, plant species composition, and N cycling feedbacks. Where information was available, we divided soil types into six textural classes for analysis: clay, clay loam, loam, silt loam, sandy loam, and sand.

For the Absolute Increase index, we performed the following Analysis of Covariance (ANCOVA) in addition to the simple regressions:

$$ANPP = constant + PPTCAT + Fertilizer\ N + PPTCAT * Fertilizer\ N + Graze + Exotics + Soil,$$

where PPTCAT is the precipitation category (300, 450, 600, 900); Fertilizer N is the actual N addition rate (continuous variable), and Graze, Exotics and Soil are all categorical variables as just described.

In the second kind of analysis for the geographic data (for the other N limitation indices), we categorized the data based on levels of fertilization, and then regressed plant production response to N against MAP within levels of fertilization. We did this with the rationale that if the degree of N limitation increased as water availability increased, we should see a positive slope of our different indices of N limitation with increasing precipitation. We used 7 different levels of fertilization: control, ≤ 3, >3–5 , >5–10, >10–20, >20–50, and >50 g $N \cdot m^{-2} \cdot yr^{-1}$. These will be referred to as FertN = 0, 3, 5, 10, 20, 50, and 100, respectively, throughout the rest of this paper. Sites in the lowest and highest categories of fertilization (FertN = 3, 50, and 100) were limited to only the dry end of the gradient. Because of this, and because grouping these sites with the other fertilization categories would lead to 10-fold variation in N-addition rates in those groups, we focus on the FertN = 0, 5, 10, and 20 categories for most of the geographical analyses, except for using the high fertilization sites in the Maximum Response analysis.

We used ANCOVA to estimate effects of other site factors on our indices of N limitation. After testing for MAP∗FertN interactions, the ANCOVA design was

$$N\ limitation\ index = constant + MAP + FertN + Graze + Exotics + Soil,$$

with MAP as a covariate, FertN as a categorical variable, and grazing, dominance by exotics, and soil type as categorical variables. Because we did not have factorial combinations for most of the variables, we did not include interaction terms.

As discussed above, mean annual temperature and its effects on water availability could also influence N limitation. For our data set, there was a significant positive correlation between site precipitation and site temperature (solid line, Figure 2(a)), so we could not include MAT directly in the ANCOVA analyses. The positive correlation oversimplifies the relationship, however. From hot, dry sites at the arid end of the gradient, temperature decreases significantly as precipitation increases up to about 600 mm MAP

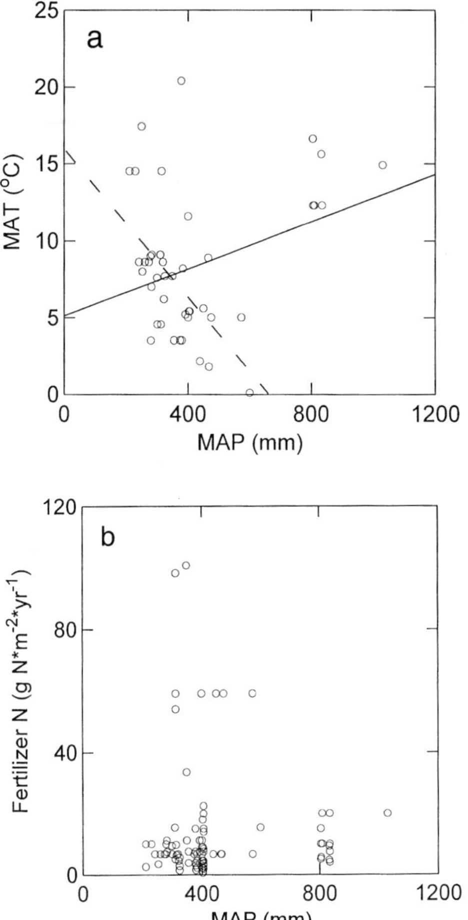

Figure 2. Distribution of sites from the literature survey in terms of (a) MAP and MAT, and (b) MAP and rate of fertilization (g N$\cdot$m$^{-2}\cdot$yr^{-1}). For (a), the dashed line is the least squared regression for all points with MAP $\leq$ 600 mm, the solid line is the regression for all points. Both correlations are significant at $P < 0.01$.

(hatched line, Figure 2(a)). Mean annual temperature increases again at the most mesic sites. To avoid the difficulty of correlation between MAP and MAT, we selected a restricted range of MAP (310–410 mm) where the N limitation indices frequently had large variance. We then did an ANCOVA similar to the one above, only using MAT instead of MAP as the covariate.

N limitation across temporal gradients in precipitation

For analyses involving year-to-year variation in precipitation, only studies with at least 4 consecutive years of data were used. We found 9 studies

(n = 129 individual data points) spanning regions of MAP from 311 mm (Lauenroth et al. 1978) to 835 mm (Konza Prairie LTER). Time spans of the studies ranged from 4 to 10 years; fertilization levels ranged from 0.6 to 22.5 g $N \cdot m^{-2} \cdot yr^{-1}$. In these analyses, we used temporal variation in actual annual precipitation (AAP, in mm) as the independent variable against which we regressed either actual ANPP, Relative Response to fertilization, or FUE. Different studies and different levels of N addition within each study were analyzed separately. We did this with the rationale that if the degree of N limitation increases as water availability increases, we should see a positive slope of our indices of N limitation with increasing AAP. We could not analyze the data using the Absolute Increase or Maximum Response indices because of lack of data.

Results

The data set

The data set includes experiments from a precipitation range from 200 to 1100 mm and a fertilization range from <1 to >100 g $N \cdot m^{-2} \cdot yr^{-1}$. However, most of the data are for levels of fertilization at or below 20 g $N \cdot m^{-2} \cdot yr^{-1}$ and within a precipitation range from 200 to 500 mm/yr (Figure 2b). Several studies had exceptionally high rates of N application ($>$ 50 g $N \cdot m^{-2} \cdot yr^{-1}$). A cluster of points around 835 mm/yr is predominantly from studies at the Konza Prairie Long-Term Ecological Research (LTER) site (see Appendix). Many of the following analyses are strongly influenced by the responses at Konza. While there are several studies from this site, we found limited information from other locations at the upper end of the precipitation gradient.

As a first evaluation of the utility of our data set, we regressed plant production in control (nonfertilized) sites against MAP. Several other studies have performed such analyses and our results compare favorably with those (Table 1). There is a large increase in ANPP with increasing precipitation, with a slope and intercept intermediate between values observed by previous studies. We did not see a saturating function as did Risser (1988), who used AET instead of MAP. This comparison gives us confidence in the utility of our data set, despite its limitations. We then compared regressions of ANPP against geographic variation in MAP for each of the fertilization levels (Table 2, Figure 3). (Note that this is not the same as our Absolute Increase index; see below). There is a significant increase in ANPP with precipitation at FertN = 10 and 20, but not FertN = 5. Only the slope for FertN = 10 is significantly greater than the slope for the nonfertilized controls. (This is

Table 1. Response of aboveground net primary productivity (ANPP) to geographic precipitation gradients. ANPP in $g \cdot m^{-2} \cdot yr^{-1}$, mean annual precipitation (MAP) in mm. MAT = Mean annual temperature. Significance < 0.001 in all cases. Equations are presented both in $y = a + bx$ form, as well as ANPP = WUE∗(MAP – IP), where WUE = average community water use efficiency (g/m^2 production per mm increase in precipitation) and IP = "ineffective precipitation", i.e., MAP at which production equals zero (Noy-Meir 1973).

Reference	Geographic range	Regression	r^2	N
This study FertN = 0	World-wide but North America-centric	ANPP = –60 + 0.49∗MAP ANPP = 0.49∗(MAP – 121)	0.66	96
(Epstein et al. 1996)	U.S. Great Plains, MAT 11.5–12 °C	ANPP = –157 + 0.76∗MAP ANPP = 0.76∗(MAP – 207)	0.87	98
(Le Houerou et al. 1988)	World-wide, but U.S.-centric	ANPP = –18.4 + 0.47∗MAP ANPP = 0.47∗(MAP – 39.4)	0.31	77
(Risser 1988)[1]	Western U.S.	ANPP = 496 – $666e^{-0.0025(ET)}$	NA	19
(Sala et al. 1988)	U.S. Great Plains	ANPP = –34 + 0.6∗MAP ANPP = 0.6∗(MAP – 56)	0.90	98
(Le Houerou & Hoste 1977)[2]	Mediterranean Basin – Europe and Africa	FU = –32.7 + 0.73∗MAP FU = 0.73∗(MAP – 45.8)	0.69	45
(Le Houerou & Hoste 1977)[2]	Sudan-Sahel	FU = 20.9 + 0.41∗MAP FU = 0.41∗(MAP + 50.9)	0.67	43
(Noy-Meir 1973)	World-wide, reported from other sources	WUE = 0.5 to 2 g/m^2 per mm IP = 25–75 mm		

[1]For linear regression of live biomass vs. MAP, $r^2 = 0.54$, but no equation is given. ET = evapotranspiration.

[2]Data presented and regressions done using units of production in Scandinavian Feed Units (FU) per ha per year. (1 FU = 1 kg barley = 1650 kcal). However, consistent conversion factors to plant production were not provided. The slope (WUE) is not comparable to the other references, but IP is. Tree and shrub productivity were not included in the Sahel data, but are probably substantial in terms of ANPP (Le Houerou & Hoste 1977).

Table 2. Response of aboveground net primary productivity (ANPP) to geographic precipitation gradients at different levels of fertilization (g N$\cdot$m$^{-2}\cdot$yr^{-1}). Equations are presented both in $y = a + bx$ form, and as parameters for the equation ANPP = WUE∗(MAP – IP). Abbreviations and units as in Table 1. Standard error of the slope (WUE) in parentheses.

Fert. Rate	$y = a + bx$	WUE (±SE)	P_{slope}	IP	$P_{int.}$	r^2	n
0	ANPP = –59.9 + 0.49∗MAP	0.49 (±0.04)	<0.001	121	0.003	0.66	96
5	ANPP = 139.1 + 0.19∗MAP	0.19 (±0.11)	0.101	–725	0.008	0.08	37
10	ANPP = –61.1 + 0.70∗MAP	0.70 (±0.08)	<0.001	86.9	0.16	0.58	57
20	ANPP = 121 + 0.40∗MAP	0.40 (±0.10)	<0.001	–305.2	0.04	0.30	38

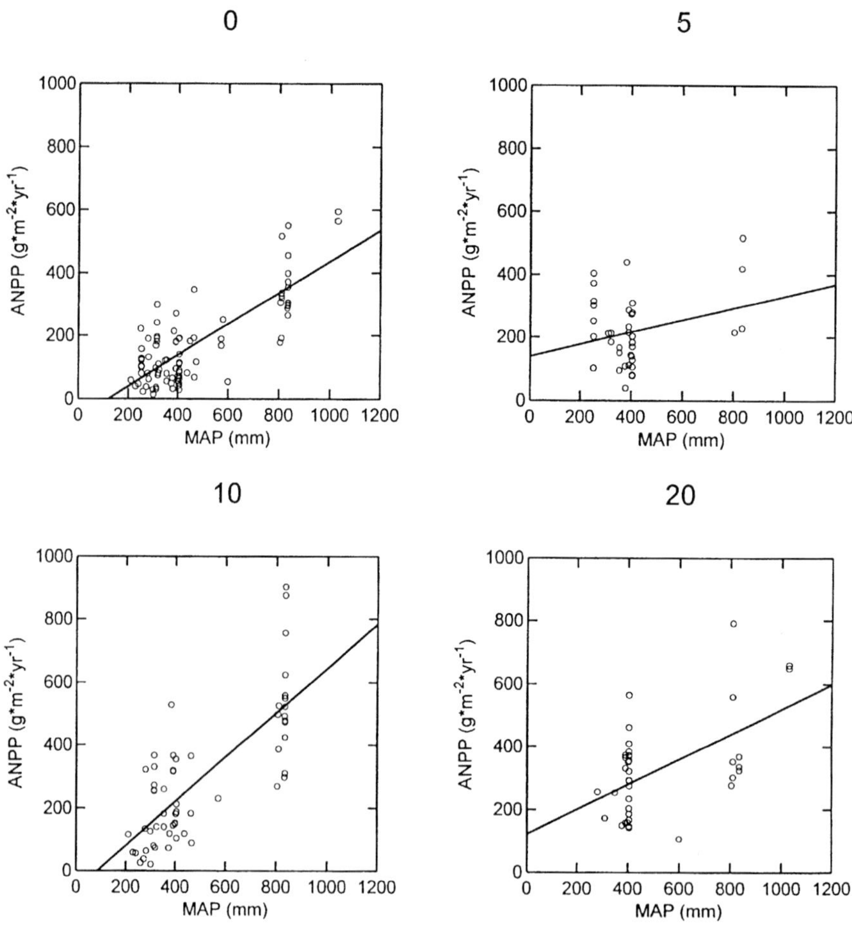

Figure 3. Response of ANPP to precipitation for different levels of FertN (see Methods).

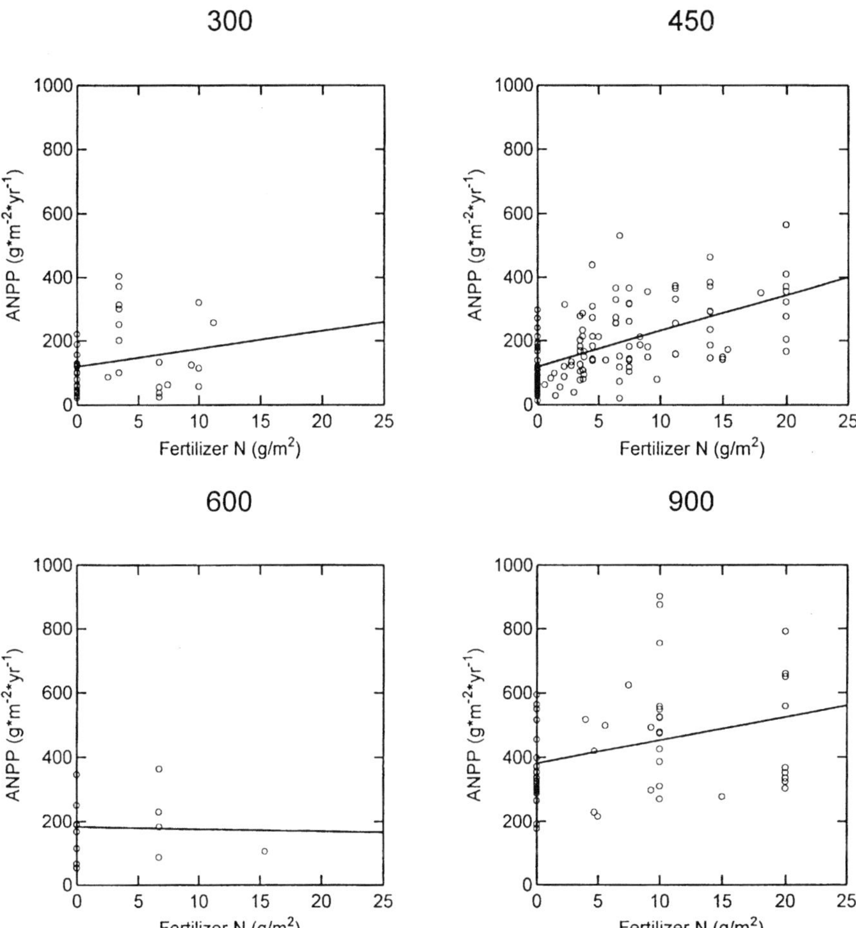

Figure 4. Absolute Increase index of N limitation: the response of ANPP to fertilization at different levels of MAP 300 is MAP $\leq$ 300 mm; 450 is MAP > 300–450 mm; 600 is MAP > 450–600 mm; 900 is MAP > 750–1100 mm. There were no data points between 600 and 750 mm MAP. Lines are least squared regressions; see Table 3 for regression parameters.

in agreement with the significant increase in fertilized ANPP with temporal variation in precipitation described later – see *ANPP response to AAP and fertilization)*. Based on this difference in slope and intercept, fertilization with 5–10 g $N \cdot m^{-2} \cdot yr^{-1}$ leads to a 50–60% increase in production all across the geographic gradient in MAP (see results for Relative Response index, below).

While substantial ancillary data exist for plant composition, grazing, soil type, and fire within certain ranges of precipitation, we rarely obtained them across the whole gradient for comparable levels of fertilization, or in factorial combinations. For example, experiments with exotics were common in some

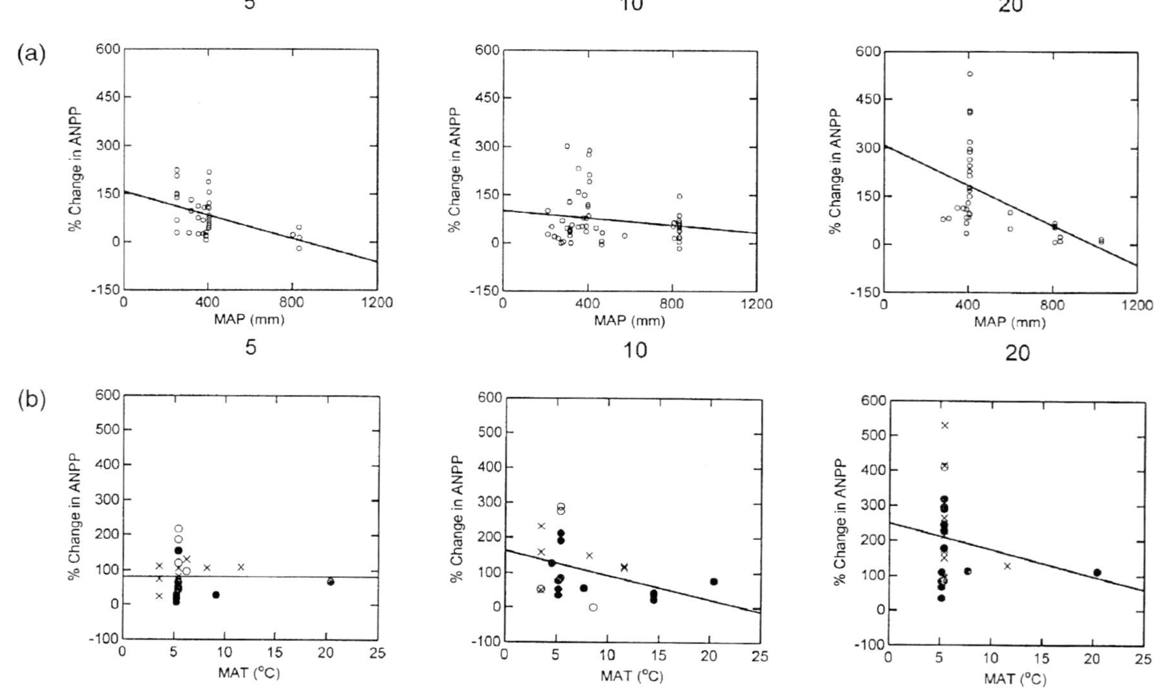

Figure 5. Relative Response of ANPP to N addition for FertN = 5, 10 and 20: (a) Relative Response as a function of MAP (see Table 4 for regression parameters); (b) Relative Response as a function of MAT for sites with MAP between 310 and 410 mm. Symbols: X = exotics, circles = natives. For natives, fill color designates grazing regime: open = ungrazed, shaded = previously grazed, filled = currently grazed.

Table 5. Response of Fertilizer Use Efficiency (FUE) to precipitation at different levels of fertilization (g N·m^{-2}·yr^{-1}). Equations are presented as FUE = a + b∗MAP, where FUE is change in ANPP /added N. Mean annual precipitation (MAP) is in mm. The regression for FertN = 10 is also presented for experiments with at least 3 years of data.

Fert. Rate	slope (±SE)	P_{slope}	intercept (±SE)	$P_{int.}$	r^2	n
5	–0.050 (±0.019)	0.012	42.7 (±8.2)	<0.001	0.17	37
10	0.018 (±0.006)	0.004	3.1 (±3.3)	0.35	0.14	57
10 (>3 yrs.)	0.031 (±0.009)	0.002	–0.1 (±4.2)	0.99	0.23	38
20	–0.013 (±0.004)	0.002	16.4 (±2.2)	<0.001	0.24	38

were much more constrained than were those at lower levels of precipitation (Figure 5), though burned sites responded more strongly than unburned sites.

We tested site MAT as a potential explanation for the large amount of variability in Relative Response among the experiments at low precipitation by regressing the data from the precipitation range of 310–410 mm against MAT (Figure 5(b)). There was no significant effect of MAT for FertN = 5 or 20. For FertN = 10, the Relative Response decreased significantly as average site temperature increased, suggesting greater N limitation in the cooler, and therefore relatively moister, sites. However, the Relative Response varied as much among experiments in the northern Great Plains (MAT ≈ 4–6 °C) as it did across the entire 5-fold range in MAT, in part due to differences in grazing regime. Sites with no grazing within the previous 5 years had highest Relative Response (ANCOVA $R^2 = 0.42$, $P_{GRAZE} = 0.017$, $n = 77$).

Fertilizer Use Efficiency Index of N limitation

Support for greater N limitation at locations with higher precipitation was equivocal in terms of Fertilizer Use Efficiency (FUE = g change in ANPP/g of fertilizer N added). In a multiple regression using all data from FertN = 5, 10, and 20, there was no significant effect of precipitation and a significant negative effect of fertilization level:

$$FUE = 22.2 - 0.0015{*}MAP - 0.745{*}\ FertN;$$

$$R^2 = 0.09,\ P_{MAP} = 0.80,\ P_{FertN} = 0.001.$$

An interaction between fertilizer category and effect of precipitation was responsible for the lack of significance of MAP in that analysis: FUE decreased with increasing MAP for FertN = 5 and 20; FUE increased with increasing MAP for FertN = 10 (Table 5, Figure 6). Plant composition, grazing regime, and soil type all had significant effects in ANCOVA but did not change the underlying interaction between MAP and FUE at different levels

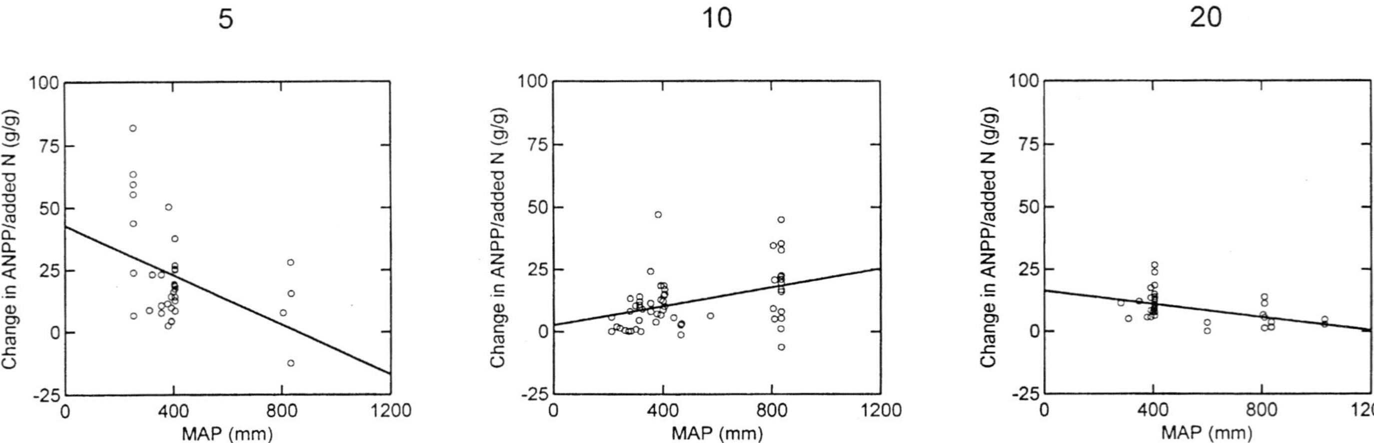

Figure 6. Response of Fertilizer Use Efficiency (change in production per amount of added N) to precipitation for FertN = 5, 10, and 20. See Table 5 for regression parameters.

of N fertilization (ANCOVA R^2 = 0.48, $P_{EXOTICS}$ = 0.003, P_{SOIL} = 0.005, P_{GRAZE} = 0.024, $P_{MAP*FERTN}$ < 0.001, n = 132). Exotics had greater responses than natives, clay soils had greater responses than sandy or loam soils, and ungrazed and currently grazed sites both had greater responses than one year previously grazed sites. The large spread among points at the lowest level of precipitation (253 mm) for FertN = 5 all came from the same site in eastern Oregon and reflect differences in response among different species of pasture grasses grown under identical conditions (Cooper 1959).

A few cases with negative values of FUE (lower production in fertilized than unfertilized plots) occurred in single year experiments in years of exceptionally low precipitation. If we constrained the analysis to only those experiments with at least three years of data to lower the possibility that the average response was unduly influenced by particularly wet or dry years, there was a stronger significant positive slope for FertN = 10 (the only category with sufficient data for this restriction; Table 5). This is in agreement with the strong production response of FUE to temporal variation in precipitation, described later.

Maximum Response Index of N limitation

The Maximum Response index of N limitation (increase in biomass at high levels of fertilization, > 20 g $N \cdot m^{-2} \cdot yr^{-1}$) tended to decrease as precipitation increased in a simple regression (Figure 7):

Change in ANPP ($g \cdot m^{-2} \cdot yr^{-1}$) = *249 – 0.17∗MAP*; P_{MAP} = 0.056, r^2 = 0.14

In ANCOVA, however, the precipitation effect was attributable to differences in species composition (ANCOVA R^2 = 0.55, P_{MAP} = 0.66, $P_{EXOTICS}$ = 0.07, n = 27). Most of the highest responses at low precipitation were from sites with exotics (*Agropyron cristatum* and *Bromus* spp.; Power 1980a; Power 1985) (Figure 7). Site MAT did not explain the variation in Maximum Response among experiments at low precipitation (MAP = 310–410 mm; ANCOVA R^2 = 0.49, P_{MAT} = 0.88, n = 15). At the mesic end of the geographic precipitation gradient, fire regime had a large effect on Maximum Response: burned tallgrass prairie accounted for the higher responses and unburned accounted for the lower responses for two studies at Konza Prairie (Seastedt et al. 1991; Benning & Seastedt 1995).

Factorial experiments with N and water additions

Another way to assess if N limitation changes relative to water limitation across the geographic precipitation gradient is to look at the response of productivity to N and water addition under different rainfall regimes. If water is predominantly limiting at the dry end and N at the wet end, then we would expect that the Relative Response to water addition would be greatest at low

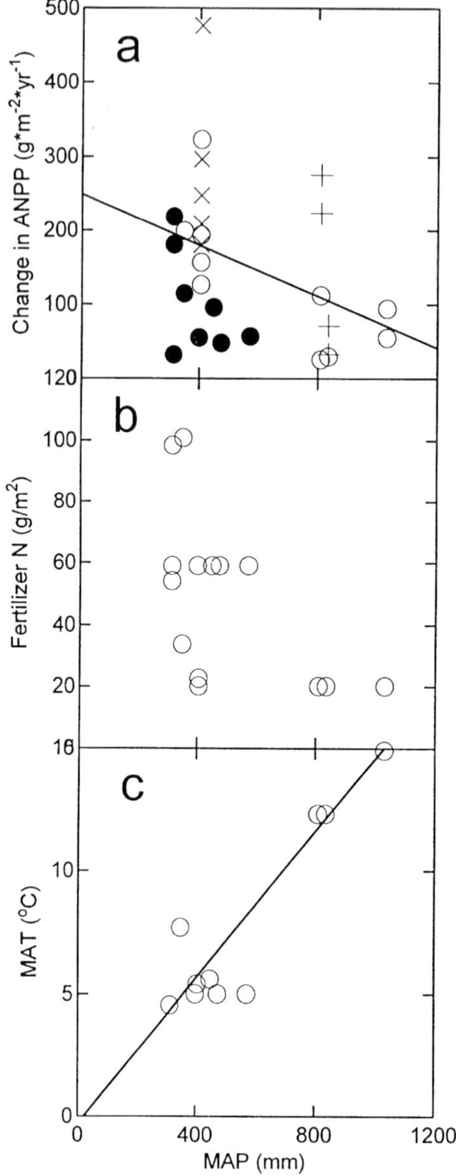

Figure 7. Maximum response of production at high levels of fertilization ($\geq$ 20 g $N \cdot m^{-2} \cdot yr^{-1}$). (a) Change in ANPP as a function of MAP. Symbols: X = exotics, + = native species, burned, and circles = native species, unburned. Shading in circles represents fertilization level: open = 20–40, gray = 50–60, and black = 90–100 g $N \cdot m^{-2} \cdot yr^{-1}$; (b) fertilization levels for studies used in the Maximum Response analysis; (c) MAT for studies used in the Maximum Response analysis.

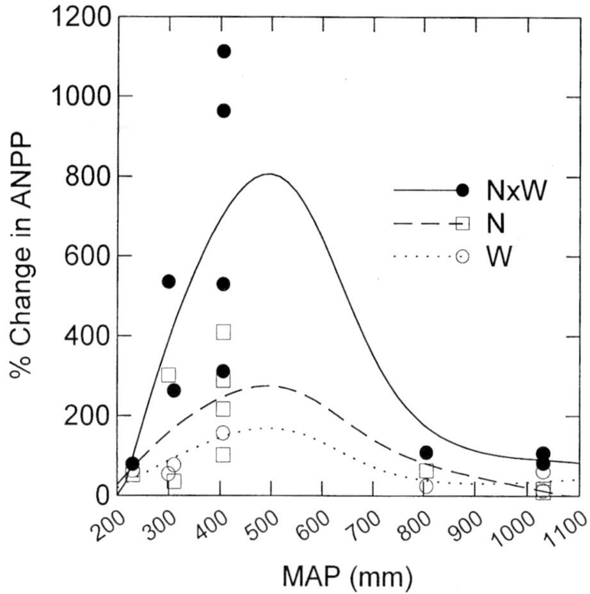

Figure 8. Relative response of ANPP to nitrogen (N), water (W), and nitrogen plus water (N × W) across the precipitation gradient for all levels of fertilization. Multiple points for N and N × W are for different levels of N addition; all studies had only one level of water addition. The curves are fitted by the distance weighted least squares procedure in SYSTAT 7.0 (SPSS 1997), and are intended as a visualization aid, not to imply a statistical relationship of response with MAP. Data are from Stephens & Whitford 1993, MAP = 230 mm; Klages & Ryerson 1965, MAP = 300 mm; Lauenroth et al. 1978, MAP = 311 mm; Smika et al. 1965, MAP = 406 mm; Owensby et al. 1970, MAP = 806 mm; and Parton & Risser 1979, MAP = 1031 mm.

precipitation and Relative Response to N addition would be greatest at high precipitation. With few exceptions, the response to N was equal to or greater than the response to water across the entire gradient (Figure 8). This held whether we looked at the Relative Response or absolute change in ANPP, and whether or not we constrained the analysis to only those cases with 3 or more years of data, or only those cases with fertilizer addition levels from 5–10 g $N \cdot m^{-2} \cdot yr^{-1}$. The greatest response to both N and water was in the middle of the precipitation gradient. N × water interactions occurred at all levels of precipitation, indicating co-limitation by both of these resources.

The only exceptions to these generalizations were at Jornada, New Mexico (230 mm; Stephens & Whitford 1993), the Central Plains Experimental Range (CPER) in Colorado (311 mm; Lauenroth et al. 1978), and in one of the grazing regimes for the modeled data from Osage, Oklahoma (≈1035 mm MAP; Parton & Risser 1979). The differences in responses to N and water for sites with low precipitation could result from differences in MAT between more southerly sites (CPER, MAT = 9.1 °C; Jornada, MAT = 14.5 °C) and more northerly sites (Norris, Montana, MAP = 300 mm, MAT = 7.6 °C; Mandan, North Dakota, MAP = 406 mm, MAT = 5.4 °C). However, MAT at CPER differs from Jornada more than it does from the northern plains sites. On the other hand, annual precipitation during the years of the experiments could have been responsible for differences among sites. For example, added N had a much stronger effect in Norris, MT than did added water, but all three years of the experiment had higher than average precipitation (Klages & Ryerson 1965). On the other hand, added water had a greater effect than did added N at CPER, but the experiment took place in years of average or below average precipitation (Lauenroth et al. 1978). The lack of long-term data spanning wet and dry years for experimental manipulations of N and water hinders our ability to explicitly test these possible explanations.

Indices of N limitation across temporal gradients in precipitation

The question we sought to answer in these analyses was whether or not N limits production more in wet years than in dry years on the same site. Within a given site, we expected a greater response to fertilization in years with higher actual annual precipitation (AAP). We also wanted to compare the magnitude of change in our N limitation indices across temporal versus geographic variation in water availability. We originally had hypothesized that the response of production to N would change more across geographic variation in precipitation than across yearly, within-site variation in precipitation.

ANPP response to AAP and fertilization

In most cases, ANPP in both control and fertilized plots was greater in wetter than drier years on the same site (Figure 9, Appendix 2). In nearly half of the experiments, AAP explains >70% of the variation in the regression with ANPP (slope $p \leq 0.1$). For several studies, AAP explains >90% of the variation in ANPP in both control and fertilized plots (slope $p < 0.05$). There are some notable exceptions, however (Power & Alessi 1971; Lauenroth et al. 1978; Konza LTER unpublished data). For these studies, some other factor besides actual annual precipitation explains variation in ANPP. Light is known to limit primary productivity due to accumulation of detritus in tallgrass prairie and fire typically decreases the degree of this limitation (Knapp & Seastedt 1986; Seastedt & Knapp 1993). Productivity in unburned plots (unburned for 10 years) at Konza Prairie showed no aboveground response to fertilizer even in years of relatively high precipitation (Figure 9, Appendix 2). In contrast, in annually burned plots at Konza, AAP

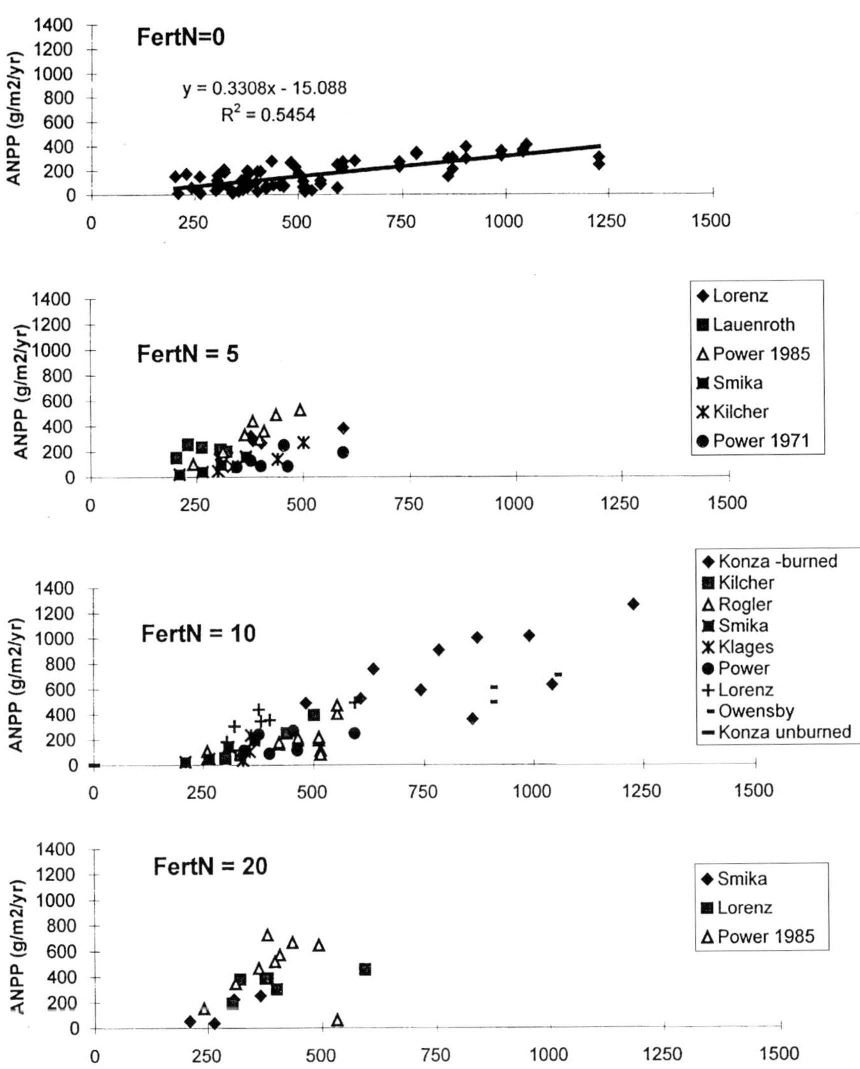

Figure 9. Response of ANPP to AAP at different levels of fertilization. One experiment in the FertN = 20 category had N addition rates of 22.5 g $N \cdot m^{-2} \cdot yr^{-1}$ (Power 1985). Regression shown for control sites only. See Appendix 2 for regression parameters for individual experiments.

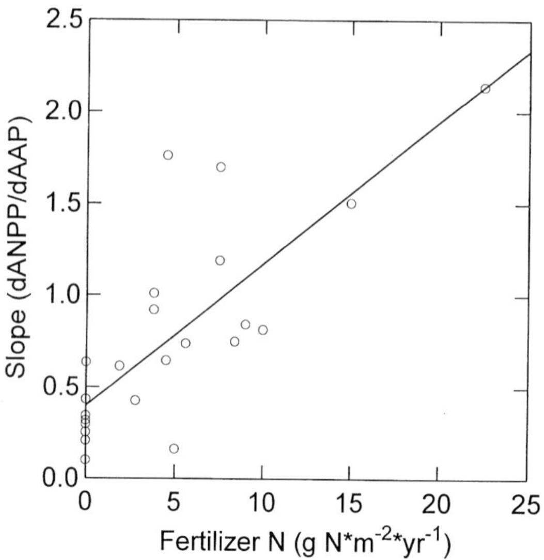

Figure 10. Effects of level of fertilization on the slope of the ANPP vs AAP regressions for within-site variation in rainfall (data from Appendix 2).

explains >40% of the variation in ANPP in fertilized plots (even more, 58%, if one outlier is removed).

Slopes in control plots (FertN = 0) range from 0.10 (Konza) to 0.64 (Kilcher 1958), with most in the range of 0.2 to 0.43. Pooling all studies, control plots had a production/precipitation slope of 0.33 ± 0.04 (Appendix 2). This regression included effects of both within- and across-site variation (since sites differ in their MAP and range of AAP). Despite this mixture of influences, the slope of this temporal relationship in control plots (FertN = 0) was still significantly less than the value for the average response of ANPP with MAP across the geographic gradient for FertN = 0 (compare Table 2). This indicates that, for control plots, geographic gradients in MAP had a larger effect on plant response to precipitation than did temporal variation in AAP, in agreement with other studies (Lauenroth & Sala 1992; Burke et al. 1997).

In fertilized plots, however, we saw the opposite. For ANPP response to precipitation at different levels of N addition, slopes of the regression lines are steeper and often more significant in the temporal (Appendix 2) than in the geographic analysis (Table 2). Furthermore, the more N added, the greater the responsiveness to increased precipitation across years on the same site (Figure 10). For all locations pooled, slopes increase from 0.33 (FertN = 0) to 0.63 (FertN = 5), to 0.74 (FertN = 10), and reach a maximum at 1.56 (FertN = 10–22.5) (Figure 9, Appendix 2). These results indicate that year to year variation in precipitation had a larger effect on plant response to N than did geographic variation in precipitation. We tested this further using the Relative Response and FUE indices of N limitation.

Relative Response Index of N limitation versus AAP

We hypothesized that the Relative Response of production to N in years of higher precipitation would be greater than in years of lower precipitation. We also predicted that the within-site slope would be lower than the between-site slope over the broad geographic gradient. Similar to the pattern across the geographic gradient with MAP, however, there was no significant effect of AAP on the Relative Response to N within any of the sites (Appendix 3). Some sites had large variations in Relative Response that was not explained by the variation in AAP (Rogler & Lorenz 1957 – with a 7-fold variation in Relative Response over a 20–25% variation in AAP; Smika et al. 1965; Power 1971). Other sites (Owensby et al. 1970; Lorenz & Rogler 1972) have a very wide range in AAP, but the same Relative Response to N across that range. We could not compare the within-site slopes to the between-sites slopes because in neither case were the regressions significant. These results suggest that the relative degree of N limitation is not related to yearly precipitation.

Fertilizer Use Efficiency Index of N limitation versus AAP

We hypothesized that FUE in fertilized plots would be greater in wetter than in drier years on the same site. This hypothesis was generally supported by the data. In approximately three-quarters of the studies, there was a significant increase in FUE with increasing AAP, indicating that plants were able to grow more per amount of added N in wetter years than in drier years (Figure 11, Appendix 4). For several studies, AAP explains >90% of the variation in FUE. Again, there are some notable exceptions (Power & Alessi 1971; Lauenroth et al. 1978).

At a given level of fertilization, slopes of FUE vs AAP within sites are greater than slopes of FUE vs MAP across sites (where slopes are negative at FertN = 5 and 20 and positive only at FertN = 10; compare Appendix 4 with Table 5). In sites with more than one level of fertilization, slopes decrease with increasing fertilization except for one study (Kilcher et al. 1965). The pattern of response to AAP within sites changes when the data are pooled across sites but within levels of fertilization (Figure 11, Appendix 4). In the latter case, the highest slopes are seen at intermediate levels of fertilization, which may occur because pooling mixes effects within sites with effects across sites. In summary, these results support our hypothesis that within a site, plants are able to use N fertilizer more efficiently (i.e. have greater gains in biomass per amount of added N) in wetter years than drier years.

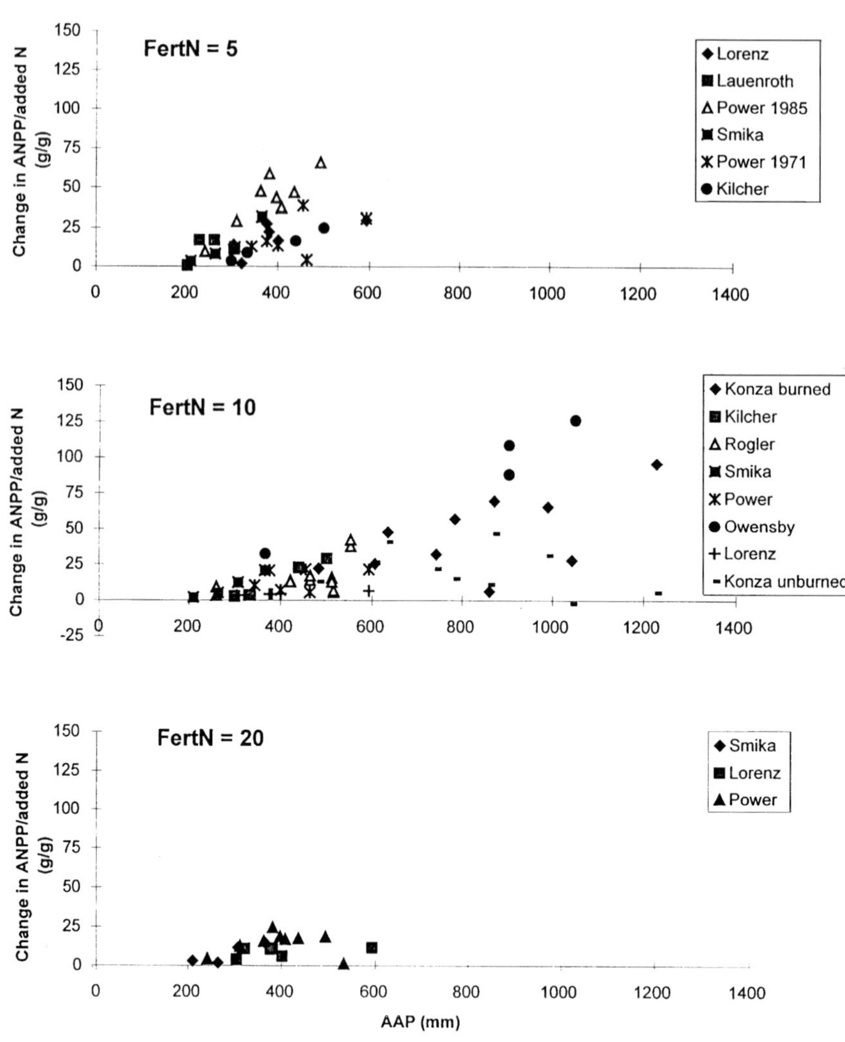

Figure 11. Response of FUE to AAP at different levels of fertilization. One experiment in the FertN = 20 category had N addition rates of 22.5 g N$\cdot$ m$^{-2}\cdot$yr^{-1} (Power 1985). See Appendix 4 for regression parameters for individual experiments.

Disscussion

Our data set shows a response of ANPP to precipitation for control sites that is similar to several other studies (Noy-Meir 1973; Le Houerou & Hoste 1977; Le Houerou et al. 1988; Risser 1988; Sala et al. 1988; Epstein et al. 1996). In general, WUE ranges from 0.46 to 0.76 g ANPP/mm MAP, and ineffective precipitation (IP) from 0 to 200 mm. We were near the lower end of the WUE range and in the middle of the IP range (Table 1). This gives us

Table 6. Summary of results of N limitation in response to variation in precipitation. Symbols are based on significant slopes: + represents positive slope, – represents negative slope, 0 = not significantly different from zero. The number of symbols indicates the strength of response (relative comparison of slopes) for those relationships with significant effects of either MAP or AAP. NA = not applicable, ND = no data.

	Geographic variation in MAP				Temporal variation in AAP			
Index	FertN 0	FertN 3–5	FertN 5–10	FertN 10–20	FertN 0	FertN 3–5	FertN 5–10	FertN 10–22.5
1. ANPP	++	++	+++	+	++	+++	+++	+++
2. Relative Response	NA	–	0	–	NA	0	0	0
3. Fertilizer Use Efficiency	NA	–	+	–	NA	+	+++	++
4. Maximum Response	NA	NA	NA	0	NA	NA	NA	ND
	MAP 300	MAP 450	MAP 600	MAP 900	MAP 300	MAP 450	MAP 600	MAP 900
5. Absolute Increase	0	++	0	+	NA	NA	NA	NA

confidence that our regression functions for control conditions are reasonably close to other studies. To our knowledge, however, this is the first attempt to look at fertilization responses across the precipitation gradient in a similar way.

The variety of responses we saw using our different indices of N limitation emphasize that the mechanisms underlying production response to fertilization need close inspection before we can infer how N limitation changes across gradients of water availability. In general, indices of N limitation based on absolute changes in plant production in response to fertilization increased with increased precipitation both temporally and geographically (Table 6). FUE increased with actual annual precipitation within sites (Appendix 4, Figure 11) and, for the intermediate level of N fertilization, FUE increased with MAP across sites (Table 5, Figure 6). Similarly, slopes of ANPP vs AAP within sites increased with fertilization, indicating alleviation of N limitation in wetter years (Appendix 2, Figure 9). Finally, Absolute Increase to fertilization was greater for MAP levels 450 and 900 mm than for 300 mm. These indices indicate that absolute production responds more to added N when water availability increases. On the other hand, regressions of FUE vs MAP at some levels of fertilization (FertN = 5 and 20), Maximum Response vs MAP, and Relative Response vs both MAP and AAP were either not significant or were negatively correlated (Table 6). These indices do not indicate that the degree of N limitation necessarily increases as water availability increases. Furthermore, production responses to N were generally greater than production responses to water addition across the precipitation gradient (Figure 8) and there were significant N $\times$ water interactions across the entire gradient, indicating co-limitation by both resources.

Either way, our results differ from the modeling analysis of Seagle and McNaughton (1993), who found water to be primarily limiting and N to be secondarily limiting in both dry and mesic regions of the Serengeti. In both regions, they saw no response to additional N supply until precipitation was at least 120% above average. In contrast, most of the studies we surveyed showed at least some response to N addition even at dry locations (Figures 5, 6, and 7) and even in years of below average precipitation (Figure 11).

Which index to use?

There is probably no one "best" measure of N limitation; each index sheds light on a different aspect of the question. Which index is best depends upon which aspect of production response to N is most relevant. For example, greater absolute increases in production in response to N addition agrees with the experience of range managers where fertilization pays for itself in wetter sites and/or wetter years in terms of increased forage production; it may not pay in drier sites or years (Kilcher 1958; Thomas and Osenbrug 1959; Seligman et al. 1986). In this sense, the Relative Response results agree with that experience – the same Relative Response across the precipitation gradient still means greater absolute production in wetter regions or years because of the greater control production under wetter conditions.

But what do these results tell us about how the degree of N limitation changes across the temporal and spatial gradients of water availability? The different responses of the different indices emphasize that this question is not as simple as it might at first appear. Our discussion will therefore focus on trying to interpret our results mechanistically to further understand the underlying interactions among N availability, water availability and plant growth. The questions we need to address are: (1) why might absolute ANPP responses to N increase with precipitation, both geographically and temporally, but Relative Response to N does not? (2) How can low precipitation sites with, on average, much lower rates of ANPP actually have equal Maximum Responses to N? (3) Why did ANPP have a higher slope in response to MAP across sites than to AAP across years in nonfertilized sites, but a higher AAP response than MAP response in fertilized sites? (4) What does each index tell us about the strength of N limitation in response to precipitation?

Several factors complicate the assessment of N limitation by using plant growth response to fertilization, especially when looking at natural ecosystems across broad gradients of environmental conditions. Our different indices help shed light on these complications. First, because the degree of microbial immobilization may change across geographic precipitation gradients (Zak et al. 1994), enough N must be added to saturate microbial demands and assure that plants are actually experiencing significantly enhanced N supply (Chapin et al. 1986; Jackson et al. 1989; Schimel et al. 1989). The Maximum Response index answers the question "No matter how much N is added, and even with changes in species, what is the greatest absolute gain in plant biomass that can be achieved?" It estimates the maximum degree to which plants can compensate for and utilize the different resources at a given site for increased biomass production. By this criterion, the degree of N limitation does not increase with increasing precipitation (Figure 7), even when taking differences in species composition (exotics vs natives) into account.

However, does one define the degree of N limitation by the Maximum Response (e.g., some index of site limitations independent of the plant community), or by the extent to which potential gains in production of the original community are limited by N supply versus other resources? A second complication is that species changes often occur with fertilization so that after multiple years of N addition, community composition can shift dramatically from the original state of the ecosystem (Tilman 1987). This is a critical question in many range studies attempting to improve grazing yields, but where weeds and unpalatable species may invade following high levels of fertilization. In many cases, species replacement in fertilization experiments follow predictable patterns, such as the increase in C3 species at the expense of C4 species in many prairie sites (Moser & Anderson 1965; Johnston et al. 1967; Goetz 1969; Rogler & Lorenz 1974), the loss of N fixers in N fertilized plots and gains in N fixers in P fertilized plots (Russel et al. 1965; Thomas & Osenbrug 1959), or the increase in ruderal species at high levels of N fertilization (Kilcher et al. 1965; Johnston et al. 1969; McLendon & Redente 1991). Such changes in species composition could be responsible for the unexpected results we saw for the Maximum Response index.

The FUE index helps to assess the extent to which production of given communities might be limited by N relative to other potentially limiting resources (specifically, water) across the precipitation gradient. The FUE index measures how absolute changes in biomass responded to fertilization at lower levels of N addition, where species shifts are less of a problem and where plant response has not yet saturated. The significant negative response of FUE to MAP at FertN = 5 (Figure 6) may result because of greater microbial immobilization in more mesic sites (resulting from higher SOM and microbial biomass), so that less added N actually reached those plants at high MAP. At FertN = 10, greater FUE at higher precipitation indicates that plants are able to convert a given amount of N into greater absolute biomass at wetter sites than at drier sites. This is similar to the positive responses seen for FUE versus AAP (see below), and in contrast to the Maximum Response index, indicates greater N limitation at sites with greater water availability. Lack of a similar response at FertN = 20 is difficult to explain.

The Relative Response to fertilization allows assessment of N response independent of initial plant size. A third complication is that despite having low nutrient supplies, plant responses to added N could be smaller on more nutrient limited sites than in more nutrient rich sites because of intrinsically low relative growth rates (RGR) of species commonly found in N limited sites (Chapin et al. 1986). If low precipitation sites had a lower absolute productivity increase in response to fertilization, does this mean that low precipitation sites are not as limited by N supply as high precipitation sites, or are the individuals just smaller and/or slower growing? Relative Response could still be influenced by differences in growth strategies if RGR differs substantially among species; however, the negative correlation of Relative Response and MAP indicates that this is not the case (see below). A similar result for Relative Response was seen in data from Johnston et al. (1969). For 5 native grassland sites in Canada ranging in MAP from ≈300 to ≈600 mm (MAT = 4.6–5.6 °C), there was a negative correlation of Relative Response to fertilizer with increasing MAP.

Mechanisms

We suggest that mechanisms at several levels could lead to no change in Relative Response or Maximum Response but significant increases in FUE as precipitation increases. Potential mechanisms include growth response to multiple resources of individual plants, plant community dynamics, and biogeochemical feedbacks at the ecosystem level. At the ecosystem level, water and N availability covary strongly across broad precipitation gradients because of linkages in their biogeochemical cycles (Schimel et al. 1997). Increasing precipitation and soil water availability positively affect atmospheric N inputs, decomposition, N mineralization, and physical transport of N ions in the soil, all of which should lead to greater plant available N (Vitousek et al. 1994; Zak et al. 1994; Burke et al. 1997; Schimel et al. 1997). Greater water availability (either geographically or temporally) could lead to greater gains in plant biomass per amount of added N (FUE) for three potential reasons: (a) more moisture makes that N more mobile and physically available in the soil (Nye & Tinker 1977), (b) plant N demand is linked to water availability (Chapin et al. 1987; Chapin 1991), or (c) individual plants are able to put acquired N into more efficient biomass production (i.e., greater carbon gain per N acquired) because of greater water availability. All of these mechanisms could potentially operate simultaneously. In the first two mechanisms, however, plants in more arid sites may be "seeing" a smaller amount of the added N, because N acquisition scales with water availability. If so, similar Relative Response to N across the geographic gradient may indicate similar capacities for species at sites of different MAP to utilize N for growth. In the last mechanism, growth response to acquired N also would be limited by water availability at low MAP. In that case, we would expect greater Relative Response of plant production to N fertilization as precipitation increases, which we did not see. Mechanisms (a) and (b) should lead to little change of plant %N with precipitation, whereas for mechanism (c), wetter sites should have lower plant %N. In a preliminary analysis of our data, fertilization up to 20 g $N \cdot m^{-2} \cdot yr^{-1}$ increased average foliar N concentrations across the entire geographical precipitation gradient, however, there were no significant effects of MAP on either the magnitude of this response or on control levels of foliar N ($P_{MAP} = 0.20$, $P_{FertN} < 0.001$, $n = 31$). In contrast to conclusions drawn by Seagle and McNaughton (1993), our results therefore suggest that mechanisms (a) and (b) are most important.

At the individual level, plants compensate for resource imbalances by allocating nonlimiting resources to the acquisition of limiting resources, with the consequence that they often are co-limited by multiple resources (Chapin et al. 1987; Chapin 1991). At the community level, those species that dominate production under a given set of conditions are presumably those that are able to convert available resources most efficiently into biomass. When changes in resource balance exceed the physiological compensation ability of the extant species, species that are able to more efficiently convert the available resources into biomass will account for a larger proportion of total production. Such changes in species composition (Lauenroth et al. 1978; Tilman 1987; Briggs & Knapp 1995) could have the effect of maintaining multiple resource limitations of the plant community as a whole. We suggest that these shifts play a large role in the patterns of ecosystem productivity in response to N that we saw across the precipitation gradient. Similarly, shifts in species dominance from year to year in response to variation in rainfall could help explain why Relative Response to fertilization did not show a significant response to AAP within sites, though more data on annual species composition is necessary to investigate this.

The lack of a geographic precipitation effect on both Relative Response to fertilization and Maximum Response to fertilization appears to negate the hypothesis that species of low resource environments have intrinsically low relative growth rates in response to addition of limiting resources (Chapin 1980; Chapin et al. 1986). Because variability in rainfall increases as MAP decreases (Noy-Meir 1973) and because plant N availability is closely tied to water availability (Chapin et al. 1987; Chapin 1991), native species in arid and semiarid systems may have the capacity to quickly respond to pulses of water and N availability, despite living in a chronically resource deficient environment. On the other hand, the original dominants of the site may indeed have low responses to added nutrients whereas the new dominants after species shifts (either invaders or previous subdominants) are better able to utilize the new balance of resources following N addition. More detailed studies that assess both production response and species changes over time across the precipitation gradient would shed more light on this.

Reasons for variation in response

Despite the strong relationship of ANPP to MAP described above, all indices of N limitation varied widely even at similar levels of MAP. This suggests that differences in site characteristics besides precipitation exert strong control over the extent to which plant growth is limited by N supply. The variation results from at least three primary factors:

a) factors that modify the relationship of plant available water to precipitation. These are primarily geophysical features, including climate (e.g., site temperature), timing and distribution of precipitation, soil texture and depth, and topography (slope and aspect).
b) factors that modify plant N availability and other resources that might limit plant growth. These include disturbance and management regimes (e.g., fire, grazing).
c) differences in plant community composition (e.g., Cooper 1959; Power 1980, 1985; Contreras & Gasto 1986).

With regard to (a), above, we had expected that MAT might help explain the variation in response to precipitation. Indeed, increasing MAT significantly decreased response to added N for the Relative Response index for FertN = 10 when looking at a restricted realm of MAP (Figure 5(b)). This is consistent with similar effects of MAT on production in unfertilized systems (Epstein et al. 1996), because increasing temperature in arid and semi-arid systems tends to decrease plant water availability. In general, however, differences among sites in other characteristics, such as grazing regime, fire regime, and plant composition influenced production response to fertilization as much as, or more than, site temperature did.

Plant production in tallgrass prairie (the primary ecosystem for which we had data at the wetter end of the gradient) is strongly affected by litter accumulation that reduces light availability and early season growing temperatures at the soil surface (Knapp & Seastedt 1986; Seastedt & Knapp 1993). Such litter feedbacks and light limitation could be one reason that we see, on average, lower Relative Response to N at higher precipitation. It is a limitation of this study that the wet end of the precipitation gradient depends so heavily on this one location and ecosystem. At the same time, this indicates that in undisturbed systems, any simple tradeoff in water and N limitation across precipitation gradients may be overwhelmed in moister regions by light limitation resulting from litter accumulation.

Production response to spatial vs temporal variation in precipitation

We originally hypothesized that interactions between N and water availability would differ for temporal versus geographic gradients in precipitation. Specifically, we expected that the production response to N addition would increase more across geographic gradients in precipitation (wet vs dry locations) than it would across temporal gradients in precipitation (i.e., wet vs dry years within a site). We predicted this based on previous studies (Lauenroth & Sala 1992; Burke et al. 1997) that found that under *control* (unfertilized) conditions, for a given increase in annual precipitation, production increased more with geographic variation in precipitation than it did with year-to-year variation in precipitation within a given site. Our literature review from control conditions (unfertilized) agrees with those findings (compare slopes for FertN = 0 in Table 2 and Appendix 2). Fertilization changes this picture, however. Temporal variation in precipitation (wet vs dry years on the same site) had a larger effect on plant response to N than did geographic variation in precipitation (wet vs dry locations) (Table 6). The slopes of ANPP vs AAP were similar to or greater than those across the geographic gradient at similar levels of N addition (compare Table 2 and Appendix 2). The same pattern holds for FUE response to geographic (Table 5) versus temporal (Appendix 4) variation in precipitation.

Together, these results indicate that (a) in control plots, productivity increases more with geographic than with temporal variation in precipitation; (b) absolute (but not relative) increases in productivity with N addition are greater in wet years than dry years, (c) absolute (but not relative) increases in productivity with N addition are greater in wet sites than dry sites (at least for FertN = 10), and (d) for a given increase in precipitation, response b is greater than response c. A couple of different explanations could cause these results.

First, for the pattern in control sites, plant production response to rainfall variation within a given location may be constrained because plants in more arid sites are intrinsically slow growing (Chapin et al. 1986; Lauenroth & Sala 1992; Burke et al. 1997). Our fertilization results, which showed large gains in biomass even in relatively dry sites, do not support this explanation.

Alternatively, production in unfertilized plots could be co-limited by both N and water because of temporal shifts in the relative availability of these resources. If water is limiting in dry years, but N is limiting in wet years, this will moderate overall production response to variation in precipitation (Van Keulen & Seligman 1992; Huston 1997). While N mineralization may increase with increasing AAP, the higher N availability may not keep pace with plant demand under greater water availability (Seagle & McNaughton 1993). With fertilization, however, N limitation is eliminated (or reduced) so that plant growth becomes singly limited by water, leading to a greater dependence of production on AAP within sites (Seligman et al. 1986). Modeling studies also show co-limitation of production by water and nitrogen when AAP is average or greater than average, indicating that alleviation of N limitation will make production more sensitive to temporal variation in annual precipitation (Van Keulen & Seligman 1992; Seagle & McNaughton 1993).

In contrast to the temporal variation, as precipitation increases across the geographic gradient, both water and N may become more available in proportion to plant demand so that co-limitation by both resources persists at different sites (Schimel et al. 1997). This could arise, for example, if greater MAP leads to greater amounts of SOM per m^2 as well as increased rates of mineralization per amount of SOM. Fertilization alleviates the N restrictions to growth, but dry years within those sites could still lower the *average* response to N (because our geographic results were means over several years of data for some sites).

In light of the modeling results that show both greater absolute and relative responses of productivity to N with increased AAP (Van Keulen & Seligman 1992; Seagle & McNaughton 1993), our results showing no such effect for Relative Response were surprising. Whereas we might have expected no relationship between Relative Response and MAP across the geographic gradient because of water and N co-limitation of production, the results from temporal variation in precipitation seem to suggest a shift from water to N limitation depending on AAP within sites. Lag effects from previous seasons' precipitation could account for some of the variability in Relative Response to N, but whether lags increase or decrease production is a topic of debate. On one hand, greater plant biomass (Lauenroth

& Sala 1992), moisture carried over from the previous year (Johnston et al. 1969), and more mineralization from previous years' senesced biomass (Burke et al. 1997) are potential mechanisms leading to greater plant growth in years following wet years. On the other hand, excess N availability in years following dry years (Risser 1988; Seligman & van Keulen 1989) or litter buildup in wet years (Burke et al. 1997) could lead to a negative correlation of previous year's precipitation with current years growth. Still other experiments report no significant carryover effects (Noy-Meir & Walker 1986). Whether positive or negative, time lags could be an important source of variation in year to year response to fertilization using current year's precipitation only. Other contributing factors to our lack of relationship between Relative Response and AAP could be that yearly shifts in species dominance within communities maintain multiple resource limitations or that other indices of water availability (e.g., growing season precipitation) are more relevant to production response to N addition than is total annual precipitation.

Conclusions

Clearly, many questions remain about how N limitation changes with water availability. We have to be cognizant of several limitations in our data set. Having information on site AET and PET would be the best way to improve our estimates of relative water availability across sites. Our survey approach did not allow us the control to closely investigate some of the substantial variation in response to N resulting from management practices that we observed. Finally, having a better representation of sites in the intermediate and mesic range of precipitation would lessen the dependence of the results on the Konza studies. This survey points out the need for further investigation, especially multi-year data sets and manipulations to investigate N × water interactions.

Despite its limitations however, we believe that this survey was able to clearly demonstrate several key points. First, mechanisms at the level of the individual, community and ecosystem all influence the response of ecosystems to changing resource availability. Second, using several indices of production response to fertilization can help give a better understanding of those mechanisms and controls of N limitation. And, third, there is not necessarily a shift of primary limitation from water to N across the geographic water availability gradient. Instead, our results support the hypothesis of co-limitation by both resources.

Acknowledgements

For motivating these questions and for discussion in the formulation of preliminary hypotheses, we would like to thank Osvaldo Sala and the other members of the Drylands group at the Termas de Chillan SCOPE N meeting: Greg Asner, Amy Austin, Tim Crews, Victor Jaramillo, Manuel Maass, Ivan Ortiz-Monasterio, Everardo Sampaio, and Eugenio Sanhueza. Several people (John Blair, Terry Chapin, Alan Knapp, Peter Vitousek, and especially Howard Epstein and an anonymous reviewer) gave extensive comments that substantially improved on an earlier draft. We would also like to thank Alan Townsend and Bob Howarth for their editorial assistance (and patience). Jesse Nippert and J.R. Matchett helped track down climate information. Unpublished data for Konza Prairie came from the LTER database. Dr. J. Krupinsky of the Northern Great Plains Research Lab provided unpublished precipitation data for Mandan, ND.

Appendix

Appendix 1. Sites and references for data used in this study. NA = data not available.

Reference	Years of experiment	Location	Country	MAP (mm)	MAT (°C)
Ettershank et al. 1978	1977	Jornada, NM	U.S.A.	211	14.5
Stephens & Whitford 1993	1988	Jornada LTER	U.S.A.	230	14.5
Kilcher et al. 1965	1958–60	Kamloops, B.C.	Canada	242	8.6
Seligman et al. 1986	1962–72	Migda	Israel	250	17.4[1]
Cooper & Hyder 1958	1957	Burns, Oregon	U.S.A.	253	8.0
Kilcher et al. 1965	1958–60	Summerland, B.C.	Canada	279	9.0
McLendon & Redente 1991	1985–89	NW Colorado	U.S.A.	280	3.5
Jacobsen et al. 1996	1971, 1977	Havre, MT	U.S.A.	282	7.0
Klipple & Retzer 1959	1953	CPER, Colorado	U.S.A.	283	9.1
Klages & Ryerson 1965	1958–60	Norris, MT	U.S.A.	300	7.6
Kilcher et al. 1965	1958–60	Manyberries, Alb.	Canada	301	4.6
Hunt et al. 1988	1984	CPER, Colorado	U.S.A.	310	9.1
Lauenroth et al. 1978	1970–74	CPER, Colorado	U.S.A.	311	9.1
Johnston et al. 1967	1964–65	Manyberries, Alb.	Canada	313	4.6
Johnston et al. 1969	1962–64	Manyberries, Alb.	Canada	313	4.6
Contreras & Gasto 1986	NA	Maipu	Chile	315	14.5[1]

Appendix 1. Continued.

Reference	Years of experiment	Location	Country	MAP (mm)	MAT (°C)
Black 1968	1961–62	Montana	U.S.A.	322	6.2
Burzlaff et al. 1968	1964	E. Montana	U.S.A.	325	7.7
Wight & Black 1979	1969	E. Montana	U.S.A.	349	7.7
Kilcher 1958	1954–57	Swift Current, Sask.	Canada	355	3.5
Kilcher et al. 1965	1958–60	Swift Current, Sask.	Canada	374	3.5
Cosper & Thomas 1961	1957–59	Texas Ag. Res. Serv., USDA	U.S.A.	378	20.4
Lodge 1959	1951	Swift Creek, Sask.	Canada	380	3.5
Thomas & Osenbrug 1959	1948–51	South Dakota	U.S.A.	383	8.2
Goetz 1969	1964–66	Havre, ND	U.S.A.	392	5.2[1]
Goetz 1969	1964–66	Manning, ND	U.S.A.	392	5.2
Goetz 1969	1964–66	Rhoades, ND	U.S.A.	392	5.2[1]
Goetz 1969	1964–66	Vebar, ND	U.S.A.	392	5.2[1]
Johnston et al. 1969	1962–64	Coalhurst, Alb.	Canada	400	5.0[1]
McGinnies 1968	1962–67	Manitou Exp. Forest, CO	U.S.A.	400	11.6
Lorenz & Rogler 1972	1958–65	Mandan, ND	U.S.A.	403	5.4
Power & Alessi 1971	1963–68	Mandan, ND	U.S.A.	403	5.4
Rogler & Lorenz 1957	1951–56	Mandan, ND	U.S.A.	406	5.4
Smika et al. 1965	1958–61	Mandan, ND	U.S.A.	406	5.4
Power 1980	1968–73	Mandan, ND	U.S.A.	406	5.4
Power 1985	1970–78	Mandan, ND	U.S.A.	406	5.4
Kilcher et al. 1965	1958–60	Indian Head, Sask.	Canada	439	2.2
Johnston et al. 1969	1962–64	Magrath, Alb.	Canada	450	5.6
Russel et al. 1965	1953	Holt County, NE	U.S.A.	465	8.9
Kilcher et al. 1965	1958–60	Brandon, Man.	Canada	468	1.8
Johnston et al. 1969	1962–64	Spring Point, Alb.	Canada	475	5.0[1]
Kilcher et al. 1965	1958–60	Stavely, Alb.	Canada	573	5.0[1]
Johnston et al. 1969	1962–64	Stavely, Alb.	Canada	573	5.0[1]
Van Keulen & Seligman 1992	1976	Niono	Mali	580	NA
Hunt et al. 1988	1984	SE Wyoming	U.S.A.	600	0.1
Elisseou et al. 1995	1992	Thessaloniki	Greece	805	16.6[1]
Owensby et al. 1970	1965–68	Flint Hills, KS	U.S.A.	806	12.3
Seastedt et al. 1991	1986–89	Konza LTER, KS	U.S.A.	810	12.3
Gay & Dwyer 1965	1963	E. Oklahoma	U.S.A.	833	15.6
Moser & Anderson 1965	1963	Flint Hills, KS	U.S.A.	835	12.3
Konza LTER (unpublished data)	1986–96	Konza LTER, KS	U.S.A.	835	12.3

Appendix 1. Continued.

Reference	Years of experiment	Location	Country	MAP (mm)	MAT (°C)
Benning & Seastedt 1995	1989–90	Konza LTER, KS	U.S.A.	835	12.3
Turner et al. 1997	1994	Konza LTER, KS	U.S.A.	835	12.3
Parton & Risser 1979	NA	Osage, OK	U.S.A.	1031	14.9

[1]MAT estimated from nearby sites or cities.

Appendix 2. Response of ANPP to temporal, within-site variation in precipitation. Regression parameters are for ANPP ($g{\cdot}m^{-2}{\cdot}yr^{-1}$) versus actual annual precipitation (AAP, in mm) within sites (ANPP = $a + b{*}$AAP). Separate regressions were performed for control plots (FertN = 0) and for each level of fertilization (FertN). Also shown are the mean annual precipitation (MAP) for each site, the range of AAP during the experiment, and the number of years for which there is data. Summary regressions for all sites within a given range of fertilization are shown at the bottom of the table. Where outliers have a large effect on parameters, regressions are performed both with and without the outliers.

Reference FertN level ($g\ N{\cdot}m^{-2}{\cdot}yr^{-1}$)	MAP (mm)	AAP range (mm)	Yrs data	r^2	slope	p	intercept	p
Lauenroth et al. 1978	311	203–501	5					
0				0.38	0.30 ± 0.22	0.26	89 ± 59	0.23
5				0.04	0.16 ± 0.44	0.74	169 ± 119	0.25
Kilcher 1958	355	300–501	4					
0				0.85	0.64 ± 0.19	0.08	–166 ± 78	0.16
3.8				0.91	1.01 ± 0.22	0.04	–262 ± 89	0.09
7.5				0.98	1.70 ± 0.17	0.01	–476 ± 69	0.02
Rogler & Lorenz 1975	406	260–516	6					
0				0.46	0.20 ± 0.07	0.02	28 ± 33	0.41
2.8				0.47	0.43 ± 0.14	0.01	–71 ± 66	0.30
8.4				0.37	0.75 ± 0.31	0.04	–142 ± 144	0.34
Smika et al. 1965	406	210–365	4					
0				0.61	0.25 ± 0.14	0.21	–45 ± 41	0.39
1.9				0.87	0.61 ± 0.17	0.07	–121 ± 49	0.13
3.8				0.95	0.92 ± 0.15	0.03	–185 ± 44	0.05
7.5				0.94	1.19 ± 0.21	0.03	–238 ± 62	0.06
15.0				0.78	1.51 ± 0.57	0.10	–290 ± 167	0.22

Appendix 2. Continued.

Reference FertN level (g N·m^{-2}·yr^{-1})	MAP (mm)	AAP range (mm)	Yrs data	r^2	slope	p	intercept	p
Lorenz & Rogler 1972	406	304–593	6					
0				0.73	0.34 ± 0.10	0.03	53 ± 43	0.28
4.5				0.77	0.65 ± 0.18	0.02	16 ± 72	0.83
9.0				0.66	0.84 ± 0.31	0.05	20 ± 124	0.87
18.0				0.44	0.59 ± 0.33	0.15	117 ± 135	0.43
Power & Alessi 1971	403	376–593	6					
0				0.23	0.12 ± 0.01	0.33	1 ± 49	0.98
0.6				0.17	0.12 ± 0.13	0.42	10 ± 59	0.87
1.1				0.09	0.11 ± 0.17	0.56	34 ± 77	0.68
2.2				0.11	0.17 ± 0.24	0.53	13 ± 109	0.91
4.5				0.30	0.43 ± 0.13	0.26	–51 ± 147	0.74
9.0				0.20	0.41 ± 0.41	0.38	1 ± 184	0.99
Power 1985	403	242–532	9					
0				0.11	0.32 ± 0.34	0.38	347 ± 59	0.0007
4.5 (w outlier)				0.06	0.49 ± 0.71	0.51	116 ± 286	0.69
4.5 (w/o outlier)				0.87	1.76 ± 0.27	0.007	–324 ± 107	0.02
22.5 (w outlier)				0.02	0.43 ± 0.99	0.68	293 ± 399	0.48
22.5 (w/o outlier)				0.75	2.14 ± 0.51	0.006	–298 ± 196	0.17
Owensby et al. 1970	806	366–1050	4					
0				0.91	0.43 ± 0.09	0.045	–43 ± 80	0.64
5.6				0.95	0.74 ± 0.12	0.03	–98 ± 104	0.45
Konza LTER	835	483–1227	10					
0 – burned (all cases)				0.43	0.10 ± 0.04	0.041	223 ± 34	0.0002
0 – burned (no outlier)				0.43	0.10 ± 0.04	0.054	224 ± 37	0.0005
10 – burned (all cases)				0.41	0.81 ± 0.34	0.047	83 ± 295	0.78
10 – burned (no outlier)				0.58	0.85 ± 0.27	0.017	99 ± 81	0.68
0 – unburned				0.02	0.51 ± 0.11	0.64	222 ± 90	0.03
10 – unburned				0.07	–1.80 ± 2.3	0.45	627 ± 198	0.01
All studies, FertN > 0 (n = 129)		203–1227		0.48	0.73 ± 0.07	0.0001	–73 ± 36	0.05

Appendix 2. Continued.

FertN level (g N·m^{-2}·yr^{-1})	r^2	slope	p	intercept	p
All sites by fertilization level:					
FertN 0 (n = 63) (excludes Konza unburned)	0.55	0.33 ± 0.04	0.0001	–15 ± 21	0.47
FertN 1–3 (n = 25)	0.33	0.30 ± 0.07	0.0003	–36 ± 32	0.27
FertN 3–5 (n = 33)	0.51	0.63 ± 0.11	0.0001	–23 ± 48	0.63
FertN 5–10 (n = 45) (excludes Konza unburned)	0.55	0.74 ± 0.08	0.0001	–52 ± 52	0.32
FertN 5–10 (no Konza outlier)	0.75	1.04 ± 0.09	0.0001	–191 ± 54	0.0009
FertN 10–22.5 (n = 18)	0.51	1.56 ± 0.40	0.0009	–206 ± 148	0.18

Appendix 3. Relative Response index of N limitation as a function of temporal variation in AAP within sites. Equations are presented as % increase in ANPP in fertilized plots = a + b∗AAP. Separate regressions are performed for each fertilizer level. Table variables are as in Appendix 2. Summary regressions for all sites within a given range of fertilization are shown at the bottom of the table. Where outliers have a large effect on parameters, regressions are performed both with and without the outliers.

Reference FertN level (g N·m^{-2}·yr^{-1})	MAP (mm)	AAP range (mm)	Yrs data	r^2	slope	p	intercept	p
Lauenroth et al. 1978	311	203–501	5					
5				0.02	–0.07 ± 0.32	0.84	46 ± 84	0.62
Kilcher 1958	355	300–501	4					
3.8				0.10	0.06 ± 0.13	0.68	36 ± 50	0.54
7.5				0.46	0.55 ± 0.42	0.32	–103 ± 17	0.60
Rogler & Lorenz 1957	406	260–516	6					
2.8				0.07	0.11 ± 0.12	0.42	40 ± 56	0.49
8.4				0.04	0.17 ± 0.26	0.54	125 ± 121	0.33
Smika et al. 1965	406	210–365	4					
1.9				0.29	0.34 ± 0.38	0.46	3 ± 110	0.98
3.8				0.10	0.57 ± 1.19	0.68	53 ± 347	0.89
7.5				0.23	0.99 ± 1.27	0.52	5 ± 373	0.99
15.0				0.08	0.48 ± 1.14	0.72	272 ± 333	0.51

Appendix 3. Continued.

Reference FertN level (g N·m^{-2}·yr^{-1})	MAP (mm)	AAP range (mm)	Yrs data	r^2	slope	p	intercept	p
Lorenz & Rogler 1972	406	304–593	6					
4.5				0.11	0.07 ± 0.09	0.52	18 ± 38	0.66
9.0				0.26	0.12 ± 0.09	0.31	40 ± 40	0.38
18.0				0.001	–0.01 ± 0.11	0.95	88 ± 43	0.11
Power 1971	403	376–593	6					
0.6				0.16	–0.07 ± 0.08	0.43	57 ± 37	0.20
1.1				0.38	–0.25 ± 0.16	0.19	177 ± 70	0.06
2.2				0.08	–0.16 ± 0.26	0.59	144 ± 118	0.29
4.5				0.002	0.04 ± 0.52	0.94	168 ± 230	0.50
9.0				0.03	–0.23 ± 0.67	0.75	377 ± 299	0.28
Power 1985	403	242–532	9					
4.5 (no effect of outlier)				0.18	–0.33 ± 0.26	0.25	250 ± 106	0.05
22.5 (no effect of outlier)				0.21	–0.62 ± 0.46	0.22	491 ± 186	0.03
Owensby et al. 1970	806	366–1050	4					
5.6				0.33	–0.02 ± 0.02	0.43	50 ± 14	0.07
Konza LTER	835	483–1227	10					
10 – burned (all cases)				0.27	0.19 ± 0.12	0.12	–17 ± 98	0.87
10 – burned (no outlier)				0.41	0.21 ± 0.10	0.06	–12 ± 81	0.88
10 – unburned				0.10	–0.09 ± 0.10	0.34	163 ± 85	0.09
All studies, all FertN (n = 129)		203–1227		0.01	–0.06 ± 0.05	0.18	163 ± 23	0.0001
All sites by fertilization level:								
FertN 1–3 (n = 33)				0.01	–0.05 ± 0.08	0.51	94 ± 33	0.008
FertN 3–5 (n = 33)				0.01	0.06 ± 0.16	0.70	86 ± 62	0.17
FertN 5–10 (n = 45)				0.07	–0.11 ± 0.05	0.06	218 ± 34	0.0001
FertN 5–10 (no Konza unburned)				0.04	–0.08 ± 0.07	0.26	233 ± 42	0.0001
FertN 10–22.5 (n = 18)				0.18	–0.67 ± 0.35	0.07	479 ± 133	0.002

Appendix 4. Response of Fertilizer Use Efficiency (FUE) to variation in actual annual precipitation (AAP) within sites. FUE is the change in ANPP/added N. Equations are presented as FUE = $a + b*$AAP. Separate regressions are performed for each fertilizer level. Table variables are as in Appendix 2. Summary regressions for all sites within a given range of fertilization are shown at the bottom of the table. Where outliers have a large effect on parameters, regressions are performed both with and without the outliers.

Reference FertN level (g N·m^{-2}·yr^{-1})	MAP (mm)	AAP range (mm)	Yrs data	r^2	slope	p	intercept	p
Lauenroth et al. 1978	311	203–501	5					
5				0.02	–0.027 ± 0.101	0.8	16.00 ± 27.1	0.59
Kilcher 1958	355	300–501	4					
3.8				0.98	0.098 ± 0.010	0.009	–25.40 ± 3.8	0.02
7.5				0.98	0.142 ± 0.015	0.01	–41.40 ± 5.9	0.02
Rogler & Lorenz 1957	406	260–516	6					
2.8				0.34	0.079 ± 0.035	0.046	–15.50 ± 16.1	0.36
8.4				0.30	0.065 ± 0.032	0.066	–13.60 ± 14.7	0.37
Smika et al. 1965	406	210-365	4					
1.9				0.87	0.191 ± 0.053	0.069	–40.40 ± 15.5	0.12
3.8				0.89	0.177 ± 0.045	0.059	–36.90 ± 13.13	0.10
7.5				0.96	0.125 ± 0.017	0.02	–25.70 ± 4.9	0.03
15.0				0.80	0.083 ± 0.030	0.11	–16.30 ± 8.7	0.20
Lorenz & Rogler 1972	406	304-593	6					
4.5				0.48	0.068 ± 0.036	0.13	–8.12 ± 14.5	0.61
9.0				0.56	0.055 ± 0.025	0.09	–3.60 ± 10.1	0.73
18.0				0.20	0.014 ± 0.014	0.37	3.59 ± 5.6	0.55
Power 1971	403	376–593	6					
0.6				0.00	–0.002 ± 0.084	0.97	19.30 ± 37.4	0.63
1.1				0.00	–0.008 ± 0.080	0.92	32.00 ± 35.5	0.41
2.2				0.02	0.022 ± 0.080	0.79	6.43 ± 35.9	0.86
4.5				0.23	0.069 ± 0.063	0.33	–11.20 ± 28.1	0.71
9.0				0.14	0.032 ± 0.040	0.469	0.38 ± 17.8	0.98
Power 1985	403	242–532	9					
4.5 (all cases)				0.02	0.038 ± 0.092	0.69	22.60 ± 37.2	0.56
4.5 (no outlier)				0.87	0.198 ± 0.046	0.005	–32.80 ± 17.8	0.11
22.5 (all cases)				0.00	0.005 ± 0.032	0.89	12.40 ± 17.7	0.36
22.5 (no outlier)				0.55	0.057 ± 0.021	0.037	–5.39 ± 8.2	0.53

Appendix 4. Continued.

Reference FertN level (g N·m^{-2}·yr^{-1})	MAP (mm)	AAP range (mm)	Yrs data	r^2	slope	p	intercept	p
Owensby et al. 1970	806	366–1050	4					
5.6				0.95	0.132 ± 0.022	0.03	–17.50 ± 18.6	0.45
Konza LTER	835	483–1227	10					
10 – burned (all cases)				0.35	0.072 ± 0.035	0.074	–14.00 ± 29.6	0.64
10 – burned (no outlier)				0.71	0.102 ± 0.010	0.0001	–25.10 ± 5.8	0.0001
10 – unburned (all cases)				0.12	–0.024 ± 0.023	0.30	40.50 ± 19.6	0.07
10 – unburned (no outlier)				0.12	–0.023 ± 0.024	0.37	40.80 ± 20.4	0.09
All studies, all fert. Combined (n = 121)		203–1227		0.49	0.083 ± 0.007	0.001	–14.50 ± 3.5	0.001
All sites by fertilization level:								
FertN 1–3 (n = 33)				0.11	0.046 ± 0.023	0.055	0.50 ± 10.1	0.96
FertN 3–5 (n = 33)				0.13	0.066 ± 0.033	0.056	–2.60 ± 13.3	0.84
FertN 5–10 (n = 45) (excludes Konza unburned)				0.64	0.092 ± 0.010	0.0001	–20.70 ± 5.3	0.0003
FertN 10–22.5 (n = 18)				0.32	0.039 ± 0.014	0.015	–2.30 ± 5.3	0.67

References

Benning TL & Seastedt TR (1995) Landscape-level interactions between topoedaphic features and nitrogen limitation in tallgrass prairie. Landscape Ecology 10: 337–348

Black AL (1968) N and phosphorus fertilization for production of crested wheatgrass and native grass in northeastern Montana. Agron. J. 60: 213–216

Blair JM (1997) Fire, nitrogen availability, and plant response in grasslands: a test of the transient maxima hypothesis. Ecol. 78: 2359–2368

Briggs JM & Knapp AK (1995) Interannual variability in primary production in tallgrass prairie: climate, soil moisture, topographic position, and fire as determinants of aboveground biomass. Am. J. Bot. 82: 1024–1030

Burke IC, Kittel TGF, Lauenroth WK, Snook P & Yonker CM (1991) Regional analysis of the central Great Plains: sensitivity to climate variability. BioSci. 41: 685–692

Burke IC, Lauenroth WK & Parton WJ (1997) Regional and temporal variation in net primary production and nitrogen mineralization in grasslands. Ecol. 78: 1330–1340

Burzlaff DF, Fick GW & Rittenhouse LR (1968) Effect of nitrogen fertilization on certain factors of a western Nebraska range ecosystem. J. Range Manage. 21: 21–23

Chapin FS, III (1980) The mineral nutrition of wild plants. Ann. Rev. Ecol. Syst. 11: 233–260

Chapin FS, III (1991) Effects of multiple environmental stresses on nutrient availability and use by plants. In: Mooney HA, Winner WE & Pell EJ (Eds) Response of Plants to Multiple Stresses (pp 67–88). Academic Press, San Diego, U.S.A.

Chapin FS, III, Bloom AJ, Field CB & Waring RH (1987) Plant responses to multiple environmental factors. BioSci. 37: 49–57

Chapin FS, III, Vitousek PM & Van Cleve K (1986) The nature of nutrient limitation in plant communities. Am. Nat. 127: 48–58

Contreras DL & Gasto JM (1986) Rangeland type and fertilization comparison in the Mediterranean part of central Chile under grazing conditions. In: Joss PJ, Lynch PW & Williams OB (Eds) Rangelands: A Resource Under Siege. Proceedings of the Second International Rangeland Congress (pp 316–317). Cambridge University Press, Cambridge

Cooper CS & Hyder DN (1958) Adaptability and yield of eleven grasses grown on the Oregon high desert. J. Range Manage. 11: 235–237

Cosper HR & Thomas JR (1961) Influence of supplemental run-off water and fertilizer on production and chemical composition of native forage. J. Range Manage. 14: 292–297

Detling JK (1979) Processes controlling blue grama production on the shortgrass prairie. In: French NR (Ed.) Perspectives in Grassland Ecology (pp 25–42). Springer-Verlag, New York, U.S.A.

Detling JK (1988) Grasslands and savannas: Regulation of energy flow and nutrient cycling by herbivores. In: Pomeroy LR & Alberts JJ (Eds) Concepts of Ecosystem Ecology (pp 131–148). Springer-Verlag, New York, U.S.A.

Elisseou GC, Veresoglou DS & Mamolos AP (1995) Vegetation productivity and diversity of acid grasslands in Northern Greece as influenced by winter rainfall and limiting nutrients. Acta Oecologica 16: 687–702

Epstein HE, Lauenroth WK & Burke IC (1997) Effects of temperature and soil texture on ANPP in the U.S. Great Plains. Ecol. 78: 2628–2631

Epstein HE, Lauenroth WK, Burke IC & Coffin DP (1996) Ecological responses of dominant grasses along two climatic gradients in the Great Plains of the United States. J. Veg. Sci. 7: 777–788

Ettershank G, Ettershank J, Bryant M & Whitford WG (1978) Effects of nitrogen fertilization on primary production in a Chihuahuan desert ecosystem. J. Arid Envir. 1: 135–139

Gay CW & Dwyer DD (1965) Effects of one year's nitrogen fertilization on native vegetation under clipping and burning. J. Range Manage. 18: 273–277

Goetz H (1969) Composition and yields of native grassland sites fertilized at different rates of nitrogen. J. Range Manage. 22: 384–390

Hunt HW, Ingham ER, Coleman DC, Elliott ET & Reid CPP (1988) Nitrogen limitation of production and decomposition in prairie, mountain meadow, and pine forest. Ecol. 69: 1009–1016

Huston, MA (1997) Hidden treatments in ecological experiments: re-evaluating the ecosystem function of biodiversity. Oecologia 110: 449–460

Jacobsen JS, Lorbeer SH, Houlton HAR & Carlson GR (1996) Nitrogen fertilization of dryland grasses in the northern great plains. J. Range Manage. 49: 340–345

Jackson LE, Schimel JP & Firestone MK (1989) Short-term partitioning of ammonium and nitrate between plants and microbes in an annual grassland. Soil Biol. and Biochem. 21: 409–415

Johnston A, Smoliak S, Smith AD & Lutwick LE (1967) Improvement of southeastern Alberta range with fertilizers. Can. J. Plant Sci. 47: 671–678

Johnston A, Smoliak S, Smith AD & Lutwick LE (1969) Seasonal precipitation, evaporation, soil moisture, and yield of fertilized range vegetation. Can. J. Plant Sci. 49: 123–128

Kilcher MR (1958) Fertilizer effects on hay production of three cultivated grasses in southern Saskatchewan. J. Range Manage. 11: 231–234

Kilcher MR, Smoliak S, Hubbard WA, Johnston A, Gross ATH & McCurdy EV (1965) Effects of inorganic nitrogen and phosphorus fertilizers on selected sites of native grassland in western Canada. Can. J. Plant Sci. 45: 229–237

Klages MG & Ryerson DE (1965) Effect of nitrogen and irrigation on yield and botanical composition of western Montana range. Agron. J.: 78–81

Klipple GE & Retzer JL (1959) Response of native vegetation of the central great plains to applications of corral manure and commercial fertilizer. J. Range Manage. 12: 239–243

Knapp AK & Seastedt TR (1986) Detritus accumulation limits production of tallgrass prairie. BioSci. 36: 662–668

Lauenroth WK, Dodd JL & Sims PL (1978) The effects of water- and nitrogen-induced stresses on plant community structure in a semiarid grassland. Oecologia 36: 211–222

Lauenroth WK & Sala OE (1992) Long-term forage production of North American shortgrass steppe. Ecol. Appl. 2: 397–403

Le Houerou HN, Bingham RL & Skerbek W (1988) Relationship between the variability of primary production and the variability of annual precipitation in world arid lands. J. Arid Environ. 15: 1–18

Le Houerou HN & Hoste CH (1977) Rangeland production and annual rainfall relations in the Mediterranean Basin and in the African Sahelo-Sudanian zone. J. Range Manage. 30: 181–189

Lodge RW (1959) Fertilization of native range in the northern great plains. J. Range Manage. 12: 277–279

Lorenz RJ & Rogler GA (1972) Forage production and botanical composition of mixed prairie as influenced by nitrogen and phosphorus fertilization. Agron. J. 64: 244–249

McGinnies WJ (1968) Effects of nitrogen fertilizer on an old stand of crested wheatgrass. Agron. J. 60: 560–562

McLendon T & Redente EF (1991) Nitrogen and phosphorus effects on secondary succession dynamics on a semi-arid sagebrush site. Ecol. 72: 2016–2024

McNaughton SJ (1985) Ecology of a grazing ecosystem: the Serengeti. Ecol. Monogr. 55: 259–294

Moser LE & Anderson KL (1965) Nitrogen and phosphorus fertilization of bluestem range. Trans. Kans. Acad. Sci. 67: 613–616

Noy-Meir I (1973) Desert ecosystems: environment and producers. Ann. Rev. Ecol. Syst. 4: 25–51

Noy-Meir I & Walker BH (1986) Stability and resilience in rangelands. In: Joss PJ, Lynch PW & Williams OB (Eds) Rangelands: A Resource under Siege. Proceedings of the Second International Rangeland Congress (pp 21–25). Cambridge University Press, Cambridge

Nye, P.H. & Tinker, P.B. 1977. Solute Movements in the Root-Soil System. Blackwell, Oxford

Owensby CE, Hyde RM & Anderson KL (1970) Effects of clipping and supplemental nitrogen and water on loamy upland bluestem range. J. Range Manage. 23: 341–346

Parton WJ & Risser PG (1979) Simulated impact of management practices upon the tallgrass prairie. In: French NR (Ed) Perspectives in Grassland Ecology (pp 135–155). Springer-Verlag, New York, U.S.A.

Power JF (1971) Evaluation of water and nitrogen stress on bromegrass growth. Agron. J. 63: 726–728

Power JF (1980a) Response of semiarid grassland sites to nitrogen fertilization: I. Plant growth and water use. Soil Sci. Soc. Am. J. 44: 545–550

Power JF (1980b) Response of semiarid grassland sites to nitrogen fertilization: II. Fertilizer recovery. Soil Sci. Soc. Am. J. 44: 550–555

Power JF (1985) Nitrogen-and water-use efficiency of several cool-season grasses receiving ammonium nitrate for 9 years. Agron. J. 77: 189–192

Power JF & Alessi J (1971) Nitrogen fertilization of semiarid grasslands: plant growth and soil mineral N levels. Agron. J. 63: 277–280

Risser PG (1988) Abiotic controls on primary productivity and nutrient cycles in North American grasslands. In: Pomeroy LR & Alberts JJ (Eds) Concepts of Ecosystem Ecology (pp 115–129). Springer-Verlag, New York, U.S.A.

Rogler GA & Lorenz RJ (1957) Nitrogen fertilization of northern great plains rangelands. J. Range Manage.: 156–160

Rogler GA & Lorenz RJ (1974) Fertilization of mid-continent range plants. In: Mays DA (Ed.) Forage Fertilization (pp 231–254). American Society of Agronomy, Crop Science Society of America, Soil Science Society of America, Madison, WI, U.S.A.

Russel JS, Brouse EM, Rhoades HF & Burzlaff DF (1965) Response of sub-irrigated meadow vegetation to application of nitrogen and phosphorus fertilizer. J. Range Manage. 18: 242–247

Sala OE, Parton WJ, Joyce LA & Lauenroth WK (1988) Primary production of the central grasslands region of the United States. Ecol. 69: 40–45

Schimel DS, Braswell BH, Holland EA, McKeown R, Ojima DS, Painter TH, Parton WJ & Townsend AR (1994) Climatic, edaphic, and biotic controls over storage and turnover of carbon in soils. Global Biogeochemical Cycles 8: 279–293

Schimel DS, VEMAP Participants & Braswell BH (1997) Continental scale variability in ecosystem processes: models, data, and the role of disturbance. Ecol. Monogr. 67: 251–271

Schimel JP, Jackson LE & Firestone MK (1989) Spatial and temporal effects on plant-microbial competition for inorganic nitrogen in a California annual grassland. Soil Biol. Biochem. 21: 1059–1066

Seagle SW & McNaughton SJ (1993) Simulated effects of precipitation and nitrogen on Serengeti grassland productivity. Biogeochemistry 22: 157–178

Seastedt TR, Briggs JM & Gibson DJ (1991) Controls of nitrogen limitation in tallgrass prairie. Oecologia 87: 72–79

Seastedt TR & Knapp AK (1993) Consequences of nonequilibrium resource availability across multiple time scales: the transient maxima hypothesis. Am. Nat. 141: 621–633

Seligman NG, Spharim I & Benjamin RW (1986) Nitrogen fertilizer application and income stability in semi-arid pastoral systems. In: Joss PJ, Lynch PW & Williams OB (Eds) Rangelands: A Resource under Siege. Proceedings of the Second International Rangeland Congress (pp 278–279). Cambridge University Press, Cambridge

Seligman NG & van Keulen H (1989) Herbage production of a Mediterranean grassland in relation to soil depth, rainfall, and nitrogen nutrition: A simulation study. Ecological Modeling 47: 303–311

Seligman NG, van Keulen H & Spitters CJT (1992) Weather, soil conditions and the inter-annual variability of herbage production and nutrient uptake on annual Mediterranean grasslands. Agricultural and Forest Meteorology 57: 265–279

Smika DE, Haas HJ & Power JF (1965) Effects of moisture and nitrogen fertilizer on growth and water use by native grass. Agron. J. 57: 483–486

SPSS Inc. (1997) SYSTAT for Windows: Graphics, Version 7.0 Edition. Chicago, Illinois, U.S.A.

Stephens G & Whitford WG (1993) Responses of *Bouteloua eriopoda* to irrigation and nitrogen fertilization in a Chihuahuan Desert grassland. J. Arid Environ. 24: 415–421

Thomas JR & Osenbrug A (1959) Effect of manure, nitrogen, phosphorus, and climatic factors on the production and quality of bromegrass-crested wheatgrass hay. Agron. J. 51: 63–66

Tilman D (1987) Secondary succession and the pattern of plant dominance along experimental nitrogen gradients. Ecol. Monogr. 57: 189–214

Turner CL, Blair JM, Schartz RJ & Neel JC (1997) Soil N and plant responses to fire, topography and supplemental N in tallgrass prairie. Ecol. 78: 1832–1843

Van Keulen H & Seligman NG (1992) Moisture, nutrient availability and plant production in the semi-arid region. In: Alberda T et al. (Eds) Food from Dry Lands (pp 25–81). Kluwer Academic, Dordrecht, The Netherlands

Vitousek PM, Aplet G, Turner D & Lockwood JJ (1992) The Mauna Loa environmental matrix: foliar and soil nutrients. Oecologia 89: 372–382

Vitousek PM, Turner D, Parton W & Sanford R (1994) Litter decomposition on the Mauna Loa Environmental Matrix, Hawaii: Patterns, mechanisms and models. Ecol. 75: 418–429

Wight JR (1976) Range fertilization in the northern great plains. J. Range Manage. 29: 180–185

Wight JR & Black AL (1979) Range fertilization: Plant response and water use. J. Range Manage. 32: 345–349

Wilkinson SR & Langdale GW (1974) Fertility needs of the warm-season grasses. In: Mays DA (Ed.) Forage Fertilization (pp 119–145). American Society of Agronomy, Crop Science Society of America, Soil Science Society of America, Madison, WI, U.S.A.

Zak DR, Tilman D, Parmenter RR, Rice CW, Fisher FM, Vose J, Milchunas D & Martin CW (1994) Plant production and soil microorganisms in late-successional ecosystems: A continental scale study. Ecol. 75: 2333–2347